The Coastlines of the World with Google Earth

Coastal Research Library

VOLUME 2

Series Editor:

Charles W. Finkl
Department of Geosciences
Florida Atlantic University
Boca Raton, FL 33431
USA

The aim of this book series is to disseminate information to the coastal research community. The Series covers all aspects of coastal research including but not limited to relevant aspects of geological sciences, biology (incl. ecology and coastal marine ecosystems), geomorphology (physical geography), climate, littoral oceanography, coastal hydraulics, environmental (resource) management, engineering, and remote sensing. Policy, coastal law, and relevant issues such as conflict resolution and risk management would also be covered by the Series. The scope of the Series is broad and with a unique crossdisciplinary nature. The Series would tend to focus on topics that are of current interest and which carry some import as opposed to traditional titles that are esoteric and non-controversial. Monographs as well as contributed volumes are welcomed.

For further volumes:
http://www.springer.com/series/8795

Anja M. Scheffers • Sander R. Scheffers • Dieter H. Kelletat

The Coastlines of the World with Google Earth

Understanding our Environment

Springer

Southern Cross University

Anja M. Scheffers
Southern Cross Geoscience
Southern Cross University
Lismore, Australia

Sander R. Scheffers
Marine Ecology Research Centre
Southern Cross University
Lismore, Australia

Dieter H. Kelletat
Department of Geography
University of Cologne
Köln, Germany

Ingeborg Scheffers, graphic design
and
Hans van der Baan, office for design

ISSN 2211-0577 e-ISSN 2211-0585
ISBN 978-94-007-0737-5 e-ISBN 978-94-007-0738-2
DOI 10.1007/978-94-007-0738-2
Springer Dordrecht Heidelberg London New York

Library of Congress Control Number: 2012932610

Printed on acid-free paper

Springer is part of Springer Science+Business Media (www.springer.com)

For Yanik and Emile

Foreword

Our living environment in all its grandeur, diversity and different scales from global to local can best be represented visually, as compared to any possible verbal descriptions. The coastlines of the world, with their total extent of at least one million kilometres, offer an excellent model for such a visualization. An overview must include the most important variations that occur in different latitudes, different geologic settings, through time and sea level changes or climatic parameters. Such detail can only be presented by the use of high resolution satellite images. Google Earth imagery mostly has such a resolution that allows visualization of individual features down to view altitudes of 500 m, which corresponds to a scale of approximately 1:5,000 or a resolution per pixel of less than 1 meter. Therefore, we chose to base this book mainly on Google Earth imagery (captured in 2010), thus showing the present day situation. Additionally, terrestrial photographs and some oblique aerial photographs taken during our various field campaigns along the coastlines of the world are added, in particular to show small features down to micro-scales and those which are hidden in vertical pictures such as steeper slopes or perpendicular cliffs. For many coastal features, information regarding their spatial extension is added in world distribution maps and more details are provided on graphs. The book shows and explains landforms and geomorphic features of different dimensions and as a result of different formational agents and processes (wind, waves, currents, tides, extreme wave events such as storms and tsunamis or by anthropogenic changes).

In our view, it is crucial for now 7 billion of us on the planet to become more knowledgeable (in contrast to being informed) about the parts and processes that currently interact on our home in the universe – planet Earth. Our ancestors have always been interested and concerned with the local weather, climate or water availability, but with the fast rate of environmental change on a global scale during the last hundred years it becomes more and more important to appreciate and understand the interconnections and interrelationships that govern Earth and create our living environment on a global scale. During the last decades, technical advances, increasing computing capacity and more sophisticated numerical modelling have transformed almost every scientific discipline into a highly complex and technical research area. Today, most scientists are specialists in one ever-narrowing research field, on the other hand the emerging science of the Earth System is changing the way scientists study Earth. With this more holistic view of the way our planet works, we want to engage, stimulate and motivate the individual person to undertake their own research and follow up with open questions in specialist texts – or even take up a career in some aspect of our Earth system.

Although this book tries to present coastal features from around the world, there are some restrictions: the low resolution or the lack of high quality pictures close to the polar regions (depending on the angle of the satellite tracks), and the difference in the spatial resolution of the images in different regions of the world, which vary from excellent (have a look at New York, where you can see single cars on the streets) to very poor, clouds may cover parts of an image, reflections from surface features (water), an unfortunate angle to the sun's rays – all these parameters may influence visibility and quality of the image data. However, Google Earth is constantly developing its set of images, and more areas will continue to appear on Google Earth's virtual globe with higher resolution in the future. Consequently, the difference in picture quality and the mosaics of different pictures from different years or with different light conditions is an obstacle in the interpretation of details. Therefore pictures of very large areas are not chosen for presentation in this book.

For this project, we would like to acknowledge the generosity of Google Earth to give permission to publish Google Earth imagery and express our thanks to Ed Parson who helped to make the book possible. As this is not a textbook for students, references are used only to acknowledge sources for material and figures, and by citing books and articles such as reviews or those presenting the state of the art for certain aspects of coastal sciences.

We gratefully acknowledge the assistance of Springer Publishing and in particular the enthusiastic support of Petra van Steenbergen, Editor of the Earth Sciences and Geography section, in preparing all parts of this book, as well as the generous support by Charles Finkl Jnr. from the Coastal Education and Research Foundation in Florida (USA) regarding copyrights of Publications in the Journal of Coastal Research. The design and layout of the book was created by Hans van der Baan and Ingeborg Scheffers in the Netherlands. Thanks to you the book is visually stunning and of high quality. Southern Cross University in Lismore (NSW, Australia) supported their work with a substantial fund based on their vision that it is vital for universities to engage with communities in ways beyond the usual academic halls. Gudrun Reichert, cartographer at the University of Duisburg-Essen (Germany) created most of the basic graphs and world maps, and Anne Hager (University of Duisburg-Essen, Germany) supported the book in many ways with her energy in problem solutions and we thank Kelly Fox (University of Queensland, Brisbane) for patience and input in text editing. We are very grateful to Bob and his passion for communicating geology. His guidance, grace and professional acumen made this book educationally sound. Finally, it should not go without saying that we are grateful to our families, our colleagues and our students at Southern Cross University (Australia) and the University of Cologne (Germany). We love working together – thank you!

Anja M. Scheffers *Sander R. Scheffers* *Dieter H. Kelletat*

About Google Earth

Virtual, web-based globes such as *Google Earth*, *NASA World Wind* or *Microsoft Virtual Earth* allow all of us to become travellers visiting the most remote places and tour our planet or even outer space at speeds faster than a rocket. Any computer user can easily, at no charge, download and use Google Earth (for both PC and Mac computers).

If you have not done so already, download *Google Earth,* version 6, from *earth.google.com,* install it on your computer, and prepare yourself to fly around the globe on your own research expedition (Fig. 1.1). You can travel to millions of locations and look for the context of all landscape features of interest to you (geography, geology, vegetation, man-made structures and more). You can also see these objects from different altitudes (i.e. in different scales), perspectives and directions; you can view a chosen area around 360° from an imaginary point in the air; and you can fly deep into canyons and craters. You can look straight down in a traditional 2D perspective or enable an oblique view in 3D, you can hover above one location, circle around or fly like a bird over countries, continents and oceans. In this book we focus on geologic and geographic features, but that is only a snapshot of what Google Earth is providing with their virtual globe. There is no room here for a complete tutorial, but you will find that the program is so easy to use and understand that you will be an expert after working with it for a few minutes. Please visit the Google Earth web page for a complete free Google Earth tutorial which is constantly updated to reflect the improvements in different versions of Google Earth (http://earth.google.com/support/bin/answer.py?hl=en&answer=176576).

We hope that the diversity of the coastlines of the world will come alive for you and stimulate your curiosity to become a coastal explorer of these fascinating places either as a hobby or profession.

Table of Contents

Introduction: Oceans and Coastlines

The most extended landforms on our planet are the coastlines; in natural scale certainly they are more than one million kilometres long. Along these vast boundaries between land and sea the variety and diversity of processes and forms leading to a coastal landscape we treasure today are immense. Some processes like the rising sea level after the melting of the large ice sheets from the last ice age are affecting coastlines globally, but some processes operate only in specific environments. Imagine for example the warm tropical waters of the lower latitudes where coral reef building organisms live and have created the largest geomorphological structure ever which can be even seen from space. These coastal environments differ completely from the ice and permafrost shaped coastlines of the Arctic regions. The coastal forms and processes we see today depend on the earlier geologic history, rock type, climatic province, sea level variations and the dynamic processes of the oceans such as waves, tides, or currents which themselves depend upon water depth, exposure, size of the ocean basin and many other factors. Coasts at the same time are regions with extreme morphological activity, comparable only with those of active plate boundaries where volcanism creates dramatic landforms or in regions where wind or ice and glaciers constantly form and sculpture the environment.

Along coastlines geology can be seen in action and you can observe forming and transforming processes even during a walk on the beach or surf in the waves. If you visit your favourite beach destination from year to year, you can trace the changes on an annual scale and often extreme events like storms change the coastline dramatically within a day or two creating new landforms or eroding large beach sections.

Whereas the surface forms under the oceans as well as those on land may be very old, from thousands of years up to tens of millions of years, all of the coastlines of the world are geologically young and represent only a tiny moment in Earth´s history, that will change dramatically in the next geological moment, and which were much different just a geologic moment ago. Sea level during the last Ice Age, about 23,000 to 18,000 years ago, was 120 m deeper than today. As the climate got warmer and the ice melted sea level reached its modern position not longer than 7,000 years ago and possibly as recently as about 6,000 years (Anthony, 2009; Bird & Schwartz, 2010; Carter, 1988; Davidson-Arnott, 2010, Kelletat, 1995; Schwartz, 2005; Woodroffe, 2003).

Scientists love to classify and categorize, seeing patterns and order in the complexity of our natural world. They invent taxonomies (Taxonomy is the art and science of classification) for plants, animals, bacteria or soils and coastal scientists are no different: They classify coastlines in attempts to characterize dominant features in terms of physical or biological properties, modes of evolution, or geographic occurrence: Is the coast advancing or retreating, emerging or submerging; do we see constructive or destructive processes operating; is the coastline rocky or sedimentary; tropical or extra-tropical; with or without sea ice; a shallow water coast or a deep water coast; exposed or sheltered; does it have high or low tide regime or it is exposed to high or low wave energy? – To give few examples!

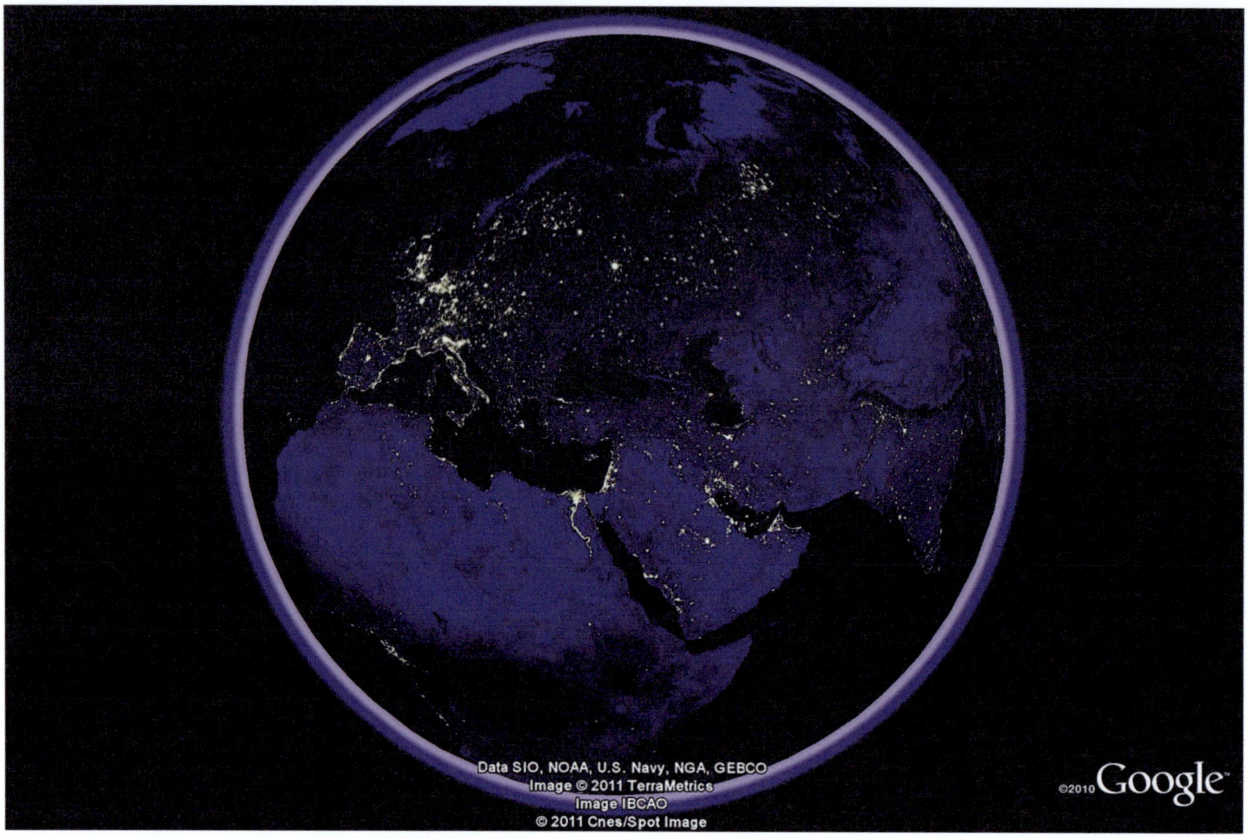

Fig. 1.1 Earth at Night (Credit: ©Google Earth). You'll find a Layers section in the sidebar. Expand the Gallery Content folder and the NASA subfolders. Then click on the small rectangles next to each option to enable it.

Another classification distinguishes between "*primary coasts*" and "*secondary coasts*". Primary coasts have preserved their initial form from terrestrial processes and now appear partly drowned by the postglacial sea-level rise. They do not show any significant transformation by coastal or marine processes since the last rise of sea level whereas the forms of secondary coasts reflect modern littoral/coastal/marine processes, mostly either by destruction (e.g. a cliff), or by construction (e.g. a beach, barrier or delta). In general, all terrestrial landforms – when partly submerged by the postglacial sea level rise – can appear as coastlines and give them their typical aspect, glacial roches moutonnés (as skerries), glacial valleys (as fjords), cone karst (as drowned karst towers), dunes and deflation depressions, river gorges (narrow rias) and many other forms. In the coastal classification system they have been given special names if they appear as coastal features.

Cities tend to grow along coastlines and transportation networks as you can see in the Night Earth view of Google Earth. Even without the underlying map, the outlines of many continents would still be visible (Fig. 1.1). They are the place where more than 45 per cent of the world's population lives and works and 75% of the mega-cities with populations over ten million are located in coastal zones. Thus, people, infrastructures and economics in coastal zones are potentially vulnerable to natural marine hazards such as storms or tsunamis as the devastating effects of Hurricane Katrina in US (2005), Cyclone Yasi (Australia, 2011), the Indian Ocean Tsunami 2004 or the powerful tsunami that hit Japan in March 2011 have shown.

Surprisingly, an unsolved question hitherto is: What is a coastline? There is no standard definition of what constitutes "the coast" because it depends largely on one's perspective or the scientific question – the coastal zone can be considered more the sea, or more the land. Imagine you have to draw the coastline of your favourite holiday destination on a map with a scale of 1:100,000. This will be easy, if there are perpendicular cliffs, but in all other cases it is difficult and needs some convention for comparison and overlap to neighbouring maps. In particular along flat depositional shorelines with high tides and storm wave impacts the actual shoreline or limit between water and land may shift for many kilometres or even tens of kilometres horizontally, at some places twice a day! If the detail of our maps is large enough (e.g. 1:10,000 to about 1:100,000), a low water coastline (MLW = mean low water) and a high water coastline (MHW = mean high water) can be differentiated, but with less detail this mostly is impossible. We can also argue that the definite limit between land and sea is along a line where sea water will never reach, but this may be far inland from the mean high water level and will differ from place to place significantly. Nevertheless these are important legal aspects, for coastal management or risk protection measures from the sea. In the following sub-chapters we will briefly present processes from the hydrosphere (the oceans), which influences the formation of coast, including organisms.

References

Anthony EJ (2009) Shore processes and their palaeoenvironmental applications. Mar Geol 4:264–288, 319–324.

Carter RWG (1988) Coastal Environments. Academic Press, London.

Davidson-Arnott R (2010) Introduction to Coastal Processes and Geomorphology. Cambridge University Press, Cambridge.

Finkl CW (2004) Coastal classification: Systematic approaches to consider in the development of a comprehensive system. Journal of Coastal Research, 20(1), 166–213.

Kelletat D H (1995) Atlas of Coastal Geomorphology and Zonality. Journal of Coastal Research Special Issue No 13.

Schwartz ML (Ed., 2005): Encyclopedia of Coastal Science. Springer, Dordrecht.

Woodroffe CD (2003) Coasts: Form, Process and Evolution. Cambridge University Press, Cambridge.

1 The Oceans

ABSTRACT As far as our present knowledge goes, Earth is unique in the solar system: It is the only planet with water in all three forms – solid, liquid and vapour – that coexist on its surface. Most of us have experienced landscapes along our vast sea shores, but in general humankind knows much less about conditions at the other side of the shoreline, in the oceans which cover 70% of Earth's surface! Before the 20th century little was known about the origin of oceans, their topography and depth or about life in them and we are only gradually coming to understand the oceans with all their geologic diversity – a fragile environment that holds a large part of Earth's biologic heritage. In the last decade, 2700 scientists from more than 50 nations participated in the "*Census of Marine Life, a Decade of Discovery*". Click on the Census of Marine Life Layer in Google Earth to learn more. The work realized by this Census, while substantial, has only scratched the surface of what remains to be learned about what lives and may live in the world's oceans. ***The Age of Discovery is not over!*** This chapter discusses our knowledge of water movement in the oceans as currents, tides and waves and we briefly overview how sea level variations have created many coastal landforms that attract us for holidays, adventure or economic reasons.

1.1 Extent, Origin and Topography

The oceans are the main features on the Earth's surface: they cover about 70.8% of it, which is nearly 362 million km^2 (Figs. 1.2–1.5 and Table 1). On the northern hemisphere, also called our land hemisphere, oceans cover 53.6% of the surface area; on the southern "water" hemisphere the vast oceans comprise 88.4% of the surface. Explore for yourself, spin the Google Earth globe and hover over both the North Pole and South Pole. This uneven distribution of land and water between both

ocean	area in Mio. km^2	content of water mass in km^3	max. depth in m	mean depth in m
Pacific Ocean	181	714,000	11022	4028
Atlantic Ocean	82,400	323,500	9219	3926
Indian Ocean	65,527	284,340	7455	3963
Arctic Ocean	14,090	13,700	5527	1205

Table 1 Dimensions for the present oceans of the world (Kelletat, 1999). We say present because the dimensions of the ocean basins change over long geologic time spans driven by plate tectonics. Also the volume of seawater in the oceans (at present 1.37 billion cubic kilometres) may change on shorter timescales over thousands of years, mainly because of the growth and melting of ice sheets and glaciers.

hemispheres plays an important role in determining the circulation in the open oceans and its marginal seas. Most of all water that exists on our planet is contained in four large interconnected basins: The Pacific Ocean, Atlantic Ocean and Indian Ocean, and the smaller Arctic Ocean. The Pacific, Atlantic and Indian Ocean are connected with the Southern Ocean, a body of water south of 60°S that encircles Antarctica and with which the Antarctic Treaty Limit coincides.

These figures include the large epicontinental seas, more or less open to their oceans like the Gulf of Mexico, the Caribbean and Hudson Bay to the Atlantic Ocean and also the Black Sea via the Mediterranean Sea, or the Persian Gulf and the Red Sea as parts of the Indian Ocean. With a total volume of about 1.37 billion km³ (which is a cube of water about 1100 × 1100 × 1100 km!) the oceans alone form 99% of the living space on our planet!

The changing geography of the oceans is due to an ever shifting pattern of tectonic plates of lithified or brittle crust (the lithosphere, Fig. 1.6) drifting on fluid rock (magma). During the last 4 billion years there were phases with super-continents and a single super-ocean, but also several break-ups into different drifting continental masses with interconnecting

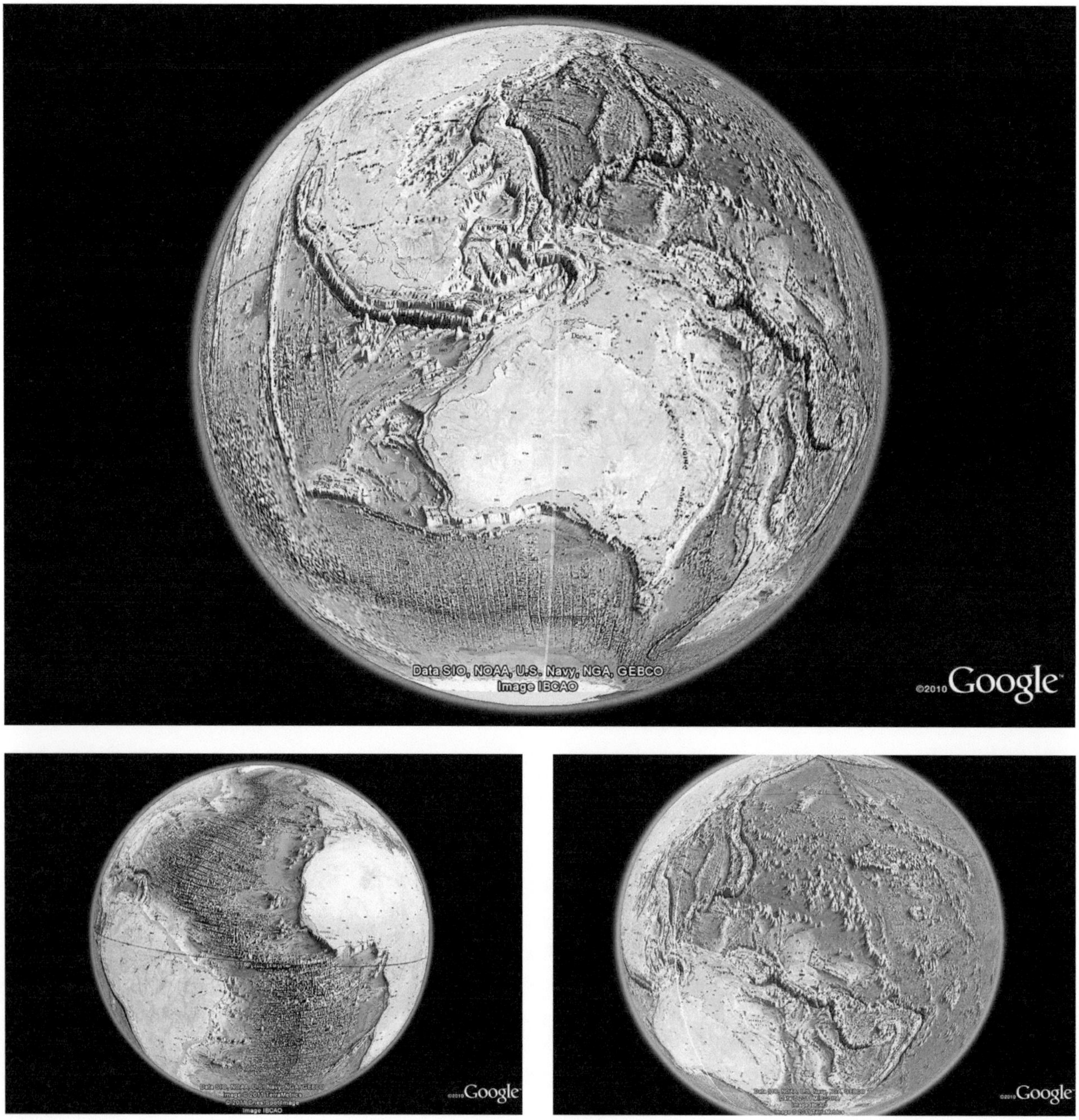

Fig. 1.2 Topography of the world's ocean floor, showing the mid ocean ridges as the main feature. This map was compiled by a path breaking team from Columbia University, i.e. Bruce Heezen and Marie Tharp. (Credit: ©Google Earth 2010: In Google Earth, click on Layers on the left, then click on Ocean; then double click on Marie Tharp Historical Map. The world will spin, and land in New York. An icon will appear for Marie Tharp Maps, LLC, double click that, and a large window opens with information on the company, as well as a link to download the map as a Google Earth layer).

seas. The formation or birth of an ocean is a rare event, yet it is unfolding today in different corners of our globe. For example, Africa is splitting apart at the seams. A plate capped by a continent, such as the African Plate, heats up from the magma below in the asthenosphere, expands and eventually splits to start a cycle of spreading. The Red Sea is a new, linear ocean that is forming where Arabia is separating and moving away from Africa. From the southern tip of the Red Sea southward through Eritrea, Ethiopia, Kenya, Tanzania and Mozambique, the African continent is rifting or splitting apart along a zone called the East African Rift. This spectacular geologic unraveling, already under way for millions of years, will be complete when saltwater from the Red Sea floods the massive rift, probably in some ten million years from now. You may also fly with Google Earth to other spreading zones, such as the Gulf of Baja California in NW Mexico.

The main process of ocean formation is called seafloor spreading. Oceanic crust of the lithosphere splits apart due to upwelling of magma and moves laterally away from the oceanic ridge. New magma rises from the mantle to fill the gap, forming new ocean crust along the ridges (Fig. 1.6). The process may unfold with a rift first appearing in a continent which becomes a graben (A *graben* is a down-thrown block which is bounded by faults along its sides) that is subsequently drowned by the ocean. A central fissure opens steadily as the lithosphere plates on both sides drift apart. Such a rift is evident in the central graben of the Mid Atlantic Ridge (Fig. 1.2). One important consequence of the spreading is that the oceanic crust far from any ridge is older than the crust nearer to the ridge.

Consider now what happens when this outpouring of basalt on the ocean bottom cools. All the minerals crystallize above 700°C – well above the point where the mineral magnetite, a component of basalt lava, crystallizes. As the lava continues to cool below 580°C, the magnetite minerals become tiny permanent magnets with the same north-south orientation (also called *polarity*) as the Earth's magnetic field at that time in history. Thus, this ancient magnetism revealed in the rocks of the ocean bottom provides a record of Earth's magnetic field of this time. Scientists studying palaeomagnetism have demonstrated that the orientation of the earth's magnetic field has frequently reversed in the Earth's geologic past resulting in parallel bands of similar magnetic properties flanking the central graben of the mid ocean ridge. This results in patterns like symmetrical bar codes reflecting the growth of the oceanic plate. The chronology of the magnetic polarity reversals can be dated very accurately so that the magnetic striping of the sea floor provides not only a tool of how old the world's oceans are, but also a means of estimating the speed with which the sea floor has

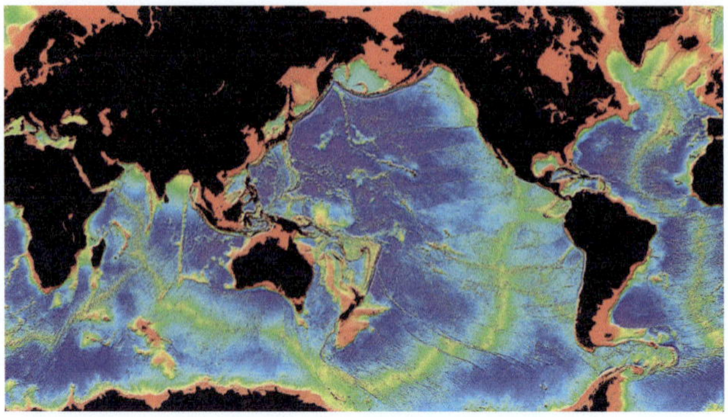

Fig. 1.3 Map of the global sea floor topography measured and estimated from gravity data using satellite altimetry and shipboard depth sounding. This map displays single undersea mountains and guyots as well as long island chains far away from the coastlines, resulting from hot spot volcanoes. (Credit: Smith, W., and Sandwell, D., 1997, Measured and Estimated Seafloor Topography, World Data Center for Geophysics & Marine Geology, Boulder Research Publication RP-1, poster, 34"×53").

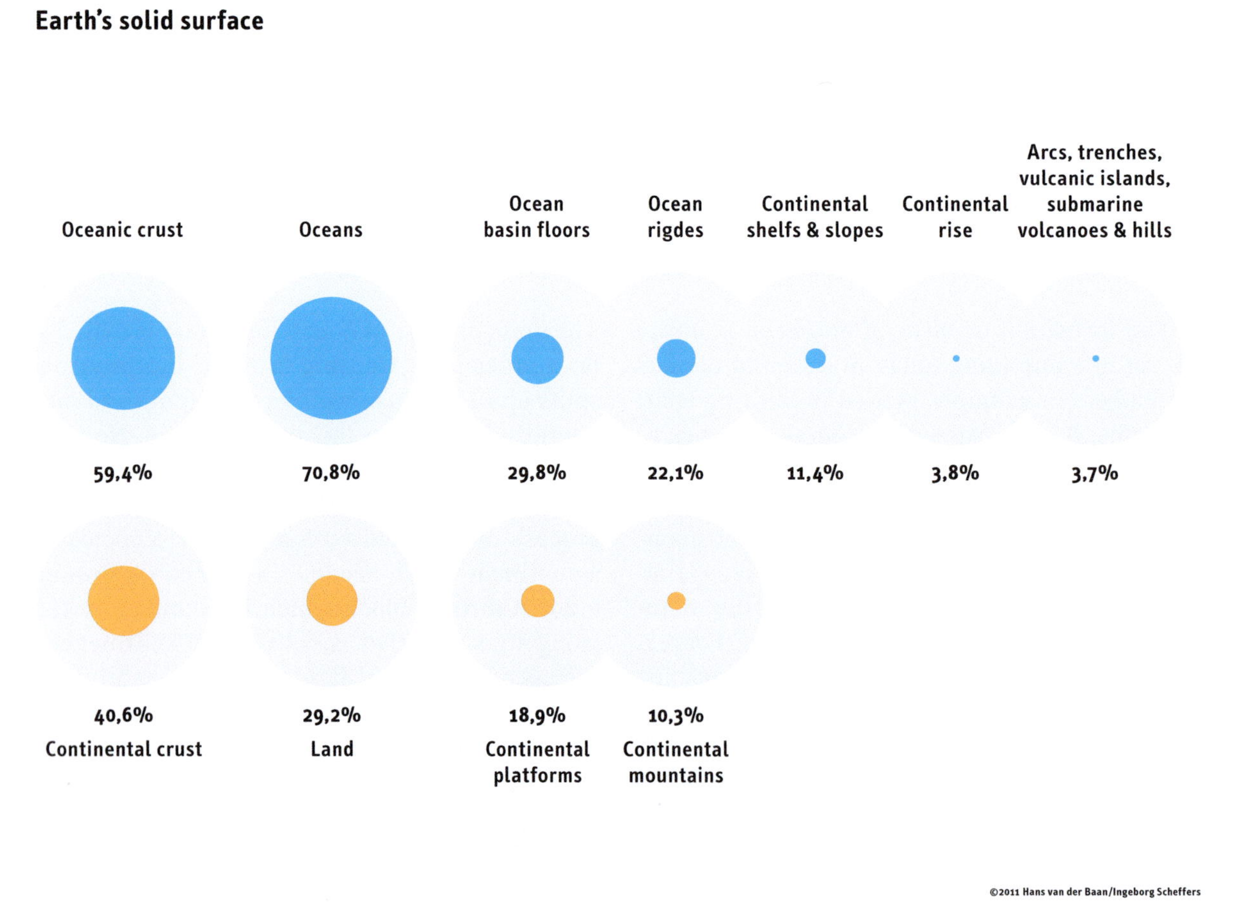

Fig. 1.4 Distribution of the main topographic features in the ocean basins.

Main forms of the ocean floor

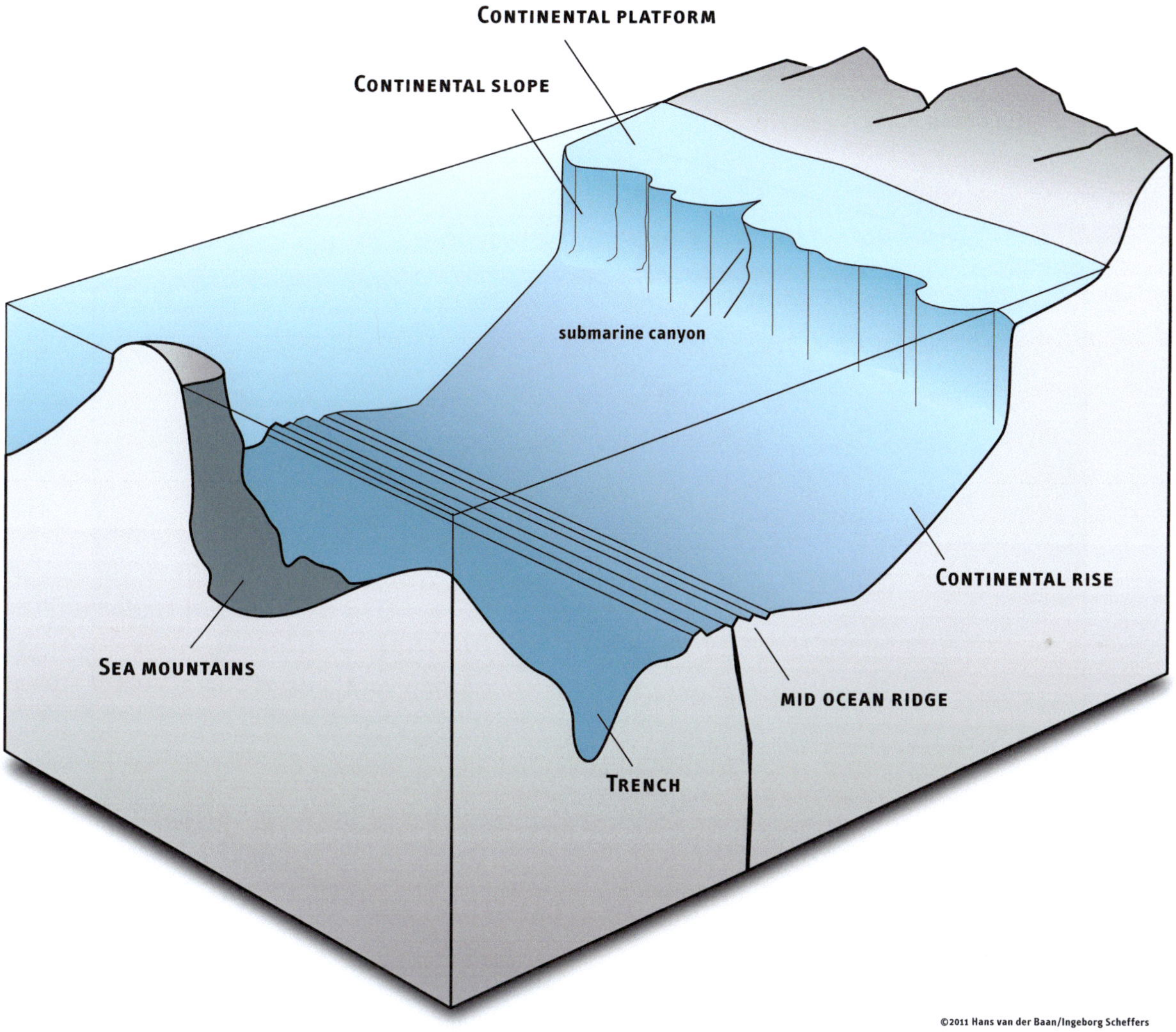

Fig. 1.5 Main features of the ocean floor.

Sea floor spreading with creation of a new ocean

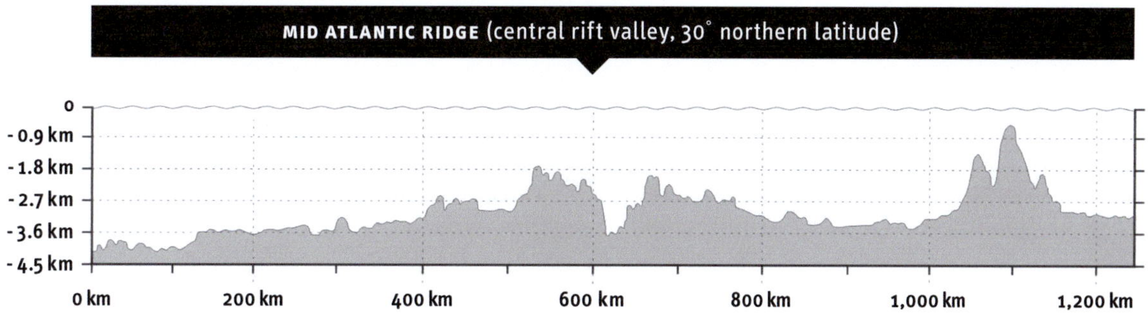

Fig. 1.6 Sea floor spreading generates new ocean floor: the rigid lithosphere drifts as plates on the asthenosphere. At the line of spreading an undersea mountain ridge (with a mid-ocean rift) is created. (Image credit: NOAA, http://sos.noaa.gov/datasets/Land/sea_floor_age.html).

Palaeomagnetic dating along central rift and mid-ocean ridges

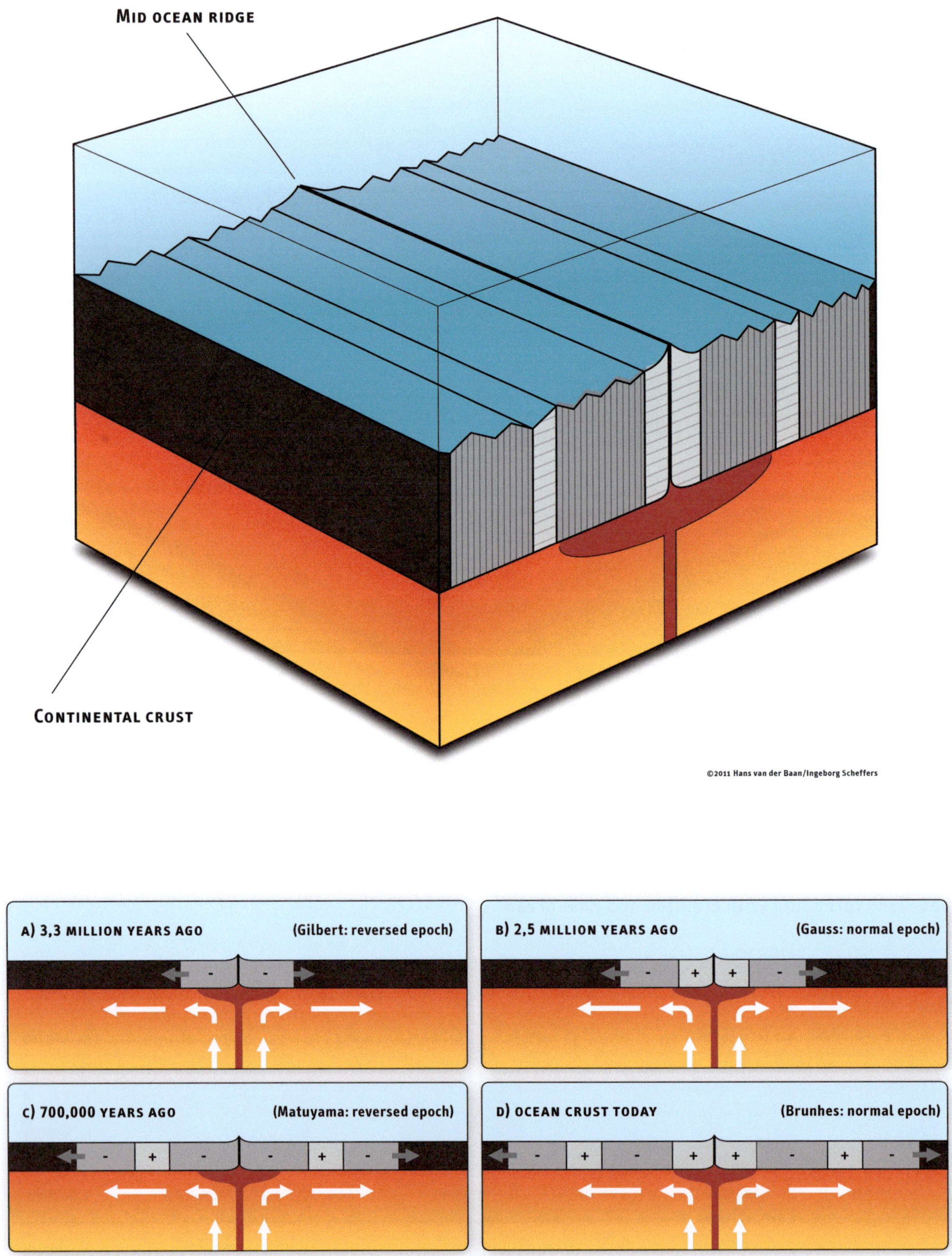

Fig. 1.7 The palaeomagnetism pattern shows symmetry in sea floor ages along both sides of the central rift of the mid-ocean ridges.

moved (Fig. 1.7). In some places, this movement is remarkably fast with velocities of 10 cm/year. It proves easy to calculate that an ocean about 4000 km wide (like the Atlantic Ocean) may have a maximum age of about 200 million years, and that the spreading rate averages several centimetres per year.

You can take a plunge beneath the surface with the Ocean Layer in Google Earth and explore what is discussed next: Along the central spreading lines in the oceans there are mid ocean ridges with a central graben. These topographic features, built by sea floor spreading, are the largest morphological element on earth, with more than 60000 km total length, relative altitudes (under water) of up to 3000 m, and widths of many hundreds of kilometres (Couper, 1983; Seibold & Berger, 1982). Iceland is one of the few places on Earth where a mid ocean ridge is above sea level (Fig. 1.8). In terms of plate tectonic framework this ridge marks the boundary between the Eurasian and North American plates. Accordingly the western part of Iceland, west of the rift zones, belongs to the North American plate and the eastern part to the Eurasian plate.

Have a look at the coastlines of Africa and South America – Their outlines are very similar, evidence that inspired the German meteorologist Alfred Wegener to formulate the continental drift hypothesis. The edges of these continents are the two sides of the rift along which the continent Gondwana first began to split and expand to form the Atlantic Ocean. However, modern shorelines most often do not coincide with the original rift because ocean water may submerge the true edges of the continents.

The flooded margin of a continent is termed the continental shelf from where a sharp drop-off (about 50 times steeper), called the continental slope, merges to the continental rise. This is an area of gently changing slope where the seafloor becomes more flat and continental crust meets oceanic crust – the true edge of the continents. The continental shelves are home to shallow sea bordering continents, mostly with very flat bottom and sloping down to water depths of about 200 m, (Fig. 1.5). The continental shelves cover about 10% of our planet and are regions where currents and waves as well as sunlight are important for inorganic and organic processes and where chang-

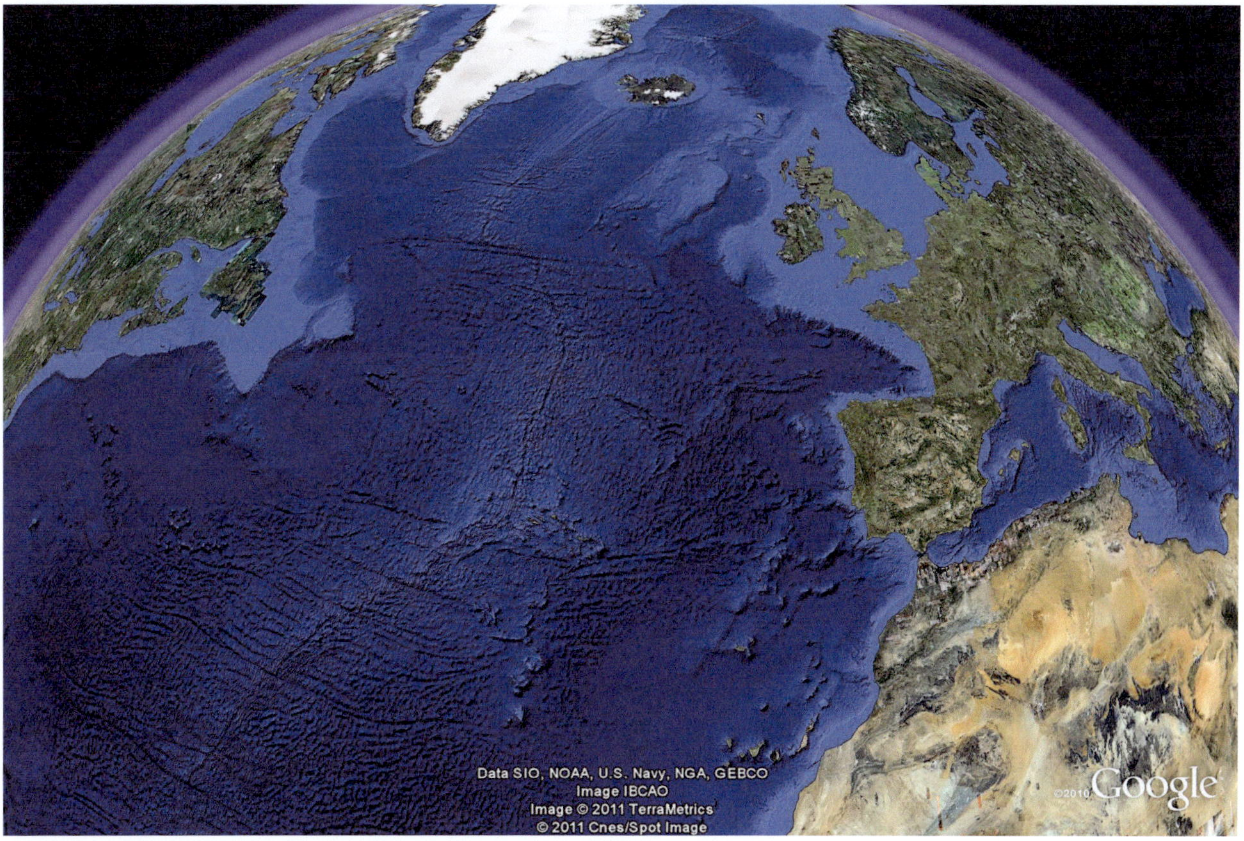

Fig. 1.8 The Mid-Atlantic Ridge, a divergent plate boundary, surfaces above sea level in Iceland (Image credit: ©Google Earth 2010).

ing sea levels have imprinted their morphological markers. On the continental slopes are valley-like features, the submarine canyons. They seem to be the result of erosion by sediment loaded flows called turbidity currents that are triggered by submarine slumps and which may gain high velocities (more than 100 km/h). Their load is dumped at the foot of the continental rise as wide sediment fans.

If ocean basins are continually changing their shapes and sizes, how does this affect the size of our planet? The process that keeps the balance is called subduction. Subduction takes place at destructive plate boundaries, where old crustal material is consumed, destroyed or we can say recycled. Here, slabs of lithosphere sink back into the asthenosphere along down-going arms of convection cells in the Earth's mantle (Figs. 1.9 and 1.10). In our modern geographic world, most of the subduction zones are located around fringes of the Pacific plate and thus, most oceanic lithosphere is destroyed in the Pacific. The Pacific is steadily getting smaller, while the Atlantic and the Indian Ocean are growing in size! Scientists have estimated that 200 million years into the future the Pacific Ocean will have disappeared and as a result, North America and Asia will collide. Subduction zones are places where immense geologic forces are evident in the form of earthquakes and active volcanoes. Around the edges of the Pacific Plate, in a zone often referred to as the Ring of Fire, about 90% of the world's earthquakes (and 81% of the world's largest) occur. The Ring of Fire is home to

Subduction under an ocean crust, forming a volcanic island

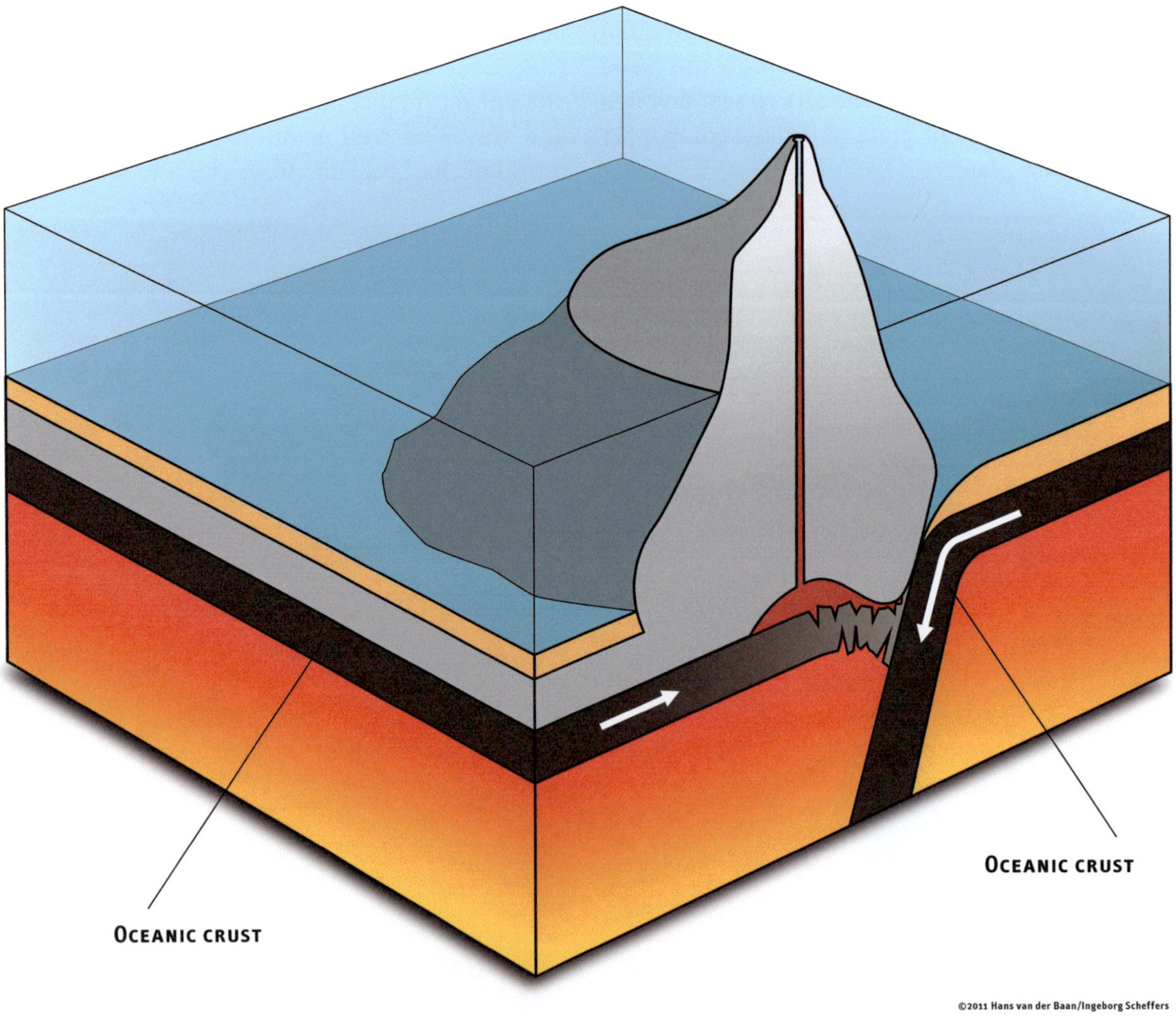

Fig. 1.9 Subduction under ocean crust, forming a volcanic island arc.

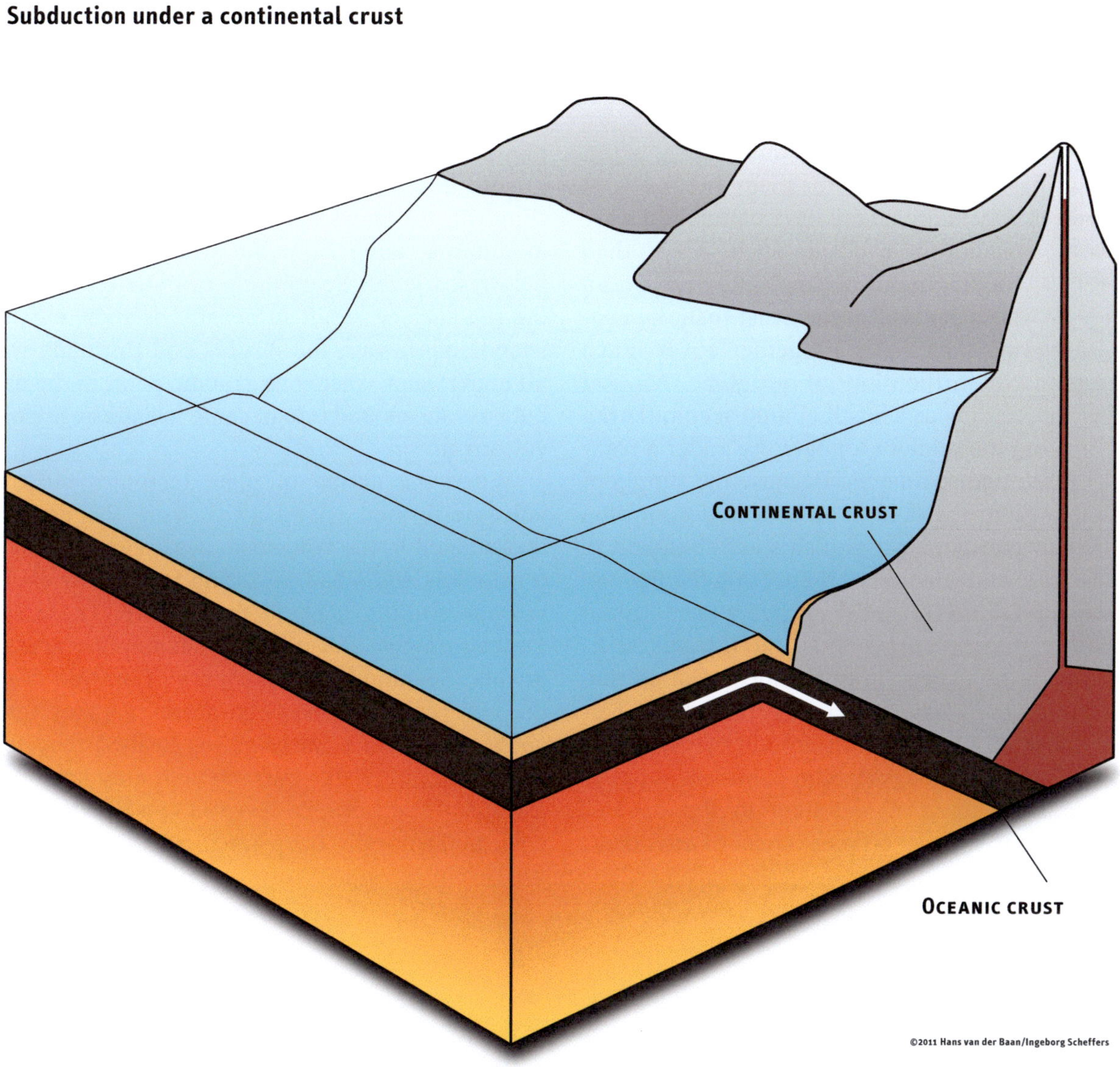

Fig. 1.10 Subduction under continental crust, forming a new mountain belt with active volcanoes.

over 75% of the world's active and dormant volcanoes. Click on *Volcanoes* in the Gallery Layer of Google Earth to learn more about the Ring of Fire.

Typical geographic features in subduction zones are deep ocean trenches that mark the places where the oceanic lithosphere grinds back into the asthenosphere. These trenches are the deepest topographic features of the world's ocean floors with the 11022 m deep Marianas Trench off the Philippines being the deepest topographic feature on Earth. On the Google Earth globe the deep trenches are clearly visible as dark, low-lying features of the ocean floor (Fig. 1.11). As an indication that subduction is an ongoing process, oceanographers observe that the deep sea trenches over subduction zones are not filled up with sediments even if they are lying in close proximity to high mountain chains. Such is the case for the trench along the Andes in the eastern Pacific Ocean.

The enormous depths of the ocean floor have been known since the days of the *Challenger* expedition in the nineteenth century. However, back then the method used to determine its depth (i.e. sending a weight attached to a line down to the ocean floor 8 km or more below) meant that only a few random measurements could be made. When these measurements were used to construct contour maps, the ocean floor looked extremely smooth. It was not until the world's ocean had been crossed many thousands of times by ships carrying echo sounders during the last decades that the ruggedness of the ocean floor was appreciated. By far the largest regions in the oceans (Seibold & Berger, 1982) are the remote deep sea basins or abyssal plains, which even today are only poorly investigated. They are several thousand metres deep and mostly without significant topography for many hundreds of kilometres. Oceanographers have claimed that we know more about the backside of the Moon than the secrets hidden in our oceans! The deep ocean basins are older parts of ocean crust than the central spreading zones and altogether cover much more than 200 million km^2 of our globe. Abyssal plains are forming as a result of mud accumulating and burying the seafloor topography under a blanket of these fine sediments. Because the supply of sediment from the land is an essential condition of their formation, abyssal plains are abundant in the Atlantic and generally absent from the Pacific, where subduction-related trenches and marginal (back-arc) basins entrap most terrigenous sediment, with the exception of its north-east corner, adjacent to the North American continent. Their importance of abyssal plains around the Antarctic is a reminder that the Antarctic continent supplies enormous volumes of sediment to the oceans, because ice is such an efficient agent of erosion.

A close look at the ocean bottom topography with Google Earth will reveal large numbers of single mountains on the ocean floor, often hundreds of kilometres across, in the vast abyssal plain. These features are called seamounts or guyots and their origin is submarine volcanism (Fig. 1.2; 1.3 and 1.5). Seamounts are undersea mountains rising from the seafloor and peaking below sea level. Seamounts tall enough to break the sea surface result in oceanic islands, e.g., the islands of Hawaii, the Azores and Bermuda were all underwater volcanoes at some point in the past, but have developed into oceanic islands by ongoing volcanism.

When the action of plate tectonics moves a volcanic-formed island away from the hot spot that created it, the volcanism ceases and the ocean crust cools and sinks, resulting in the now extinct volcano sinking beneath the surface. These submerged, often flat-topped, seamounts are the guyots. Seamounts are hotspots of marine life in the vast realms of the open ocean. As they tower above the surface surrounding seabed they tend to concentrate water currents and may have their own localised tides, eddies and upwellings (where cold, nutrient-rich, deepwater moves up along the

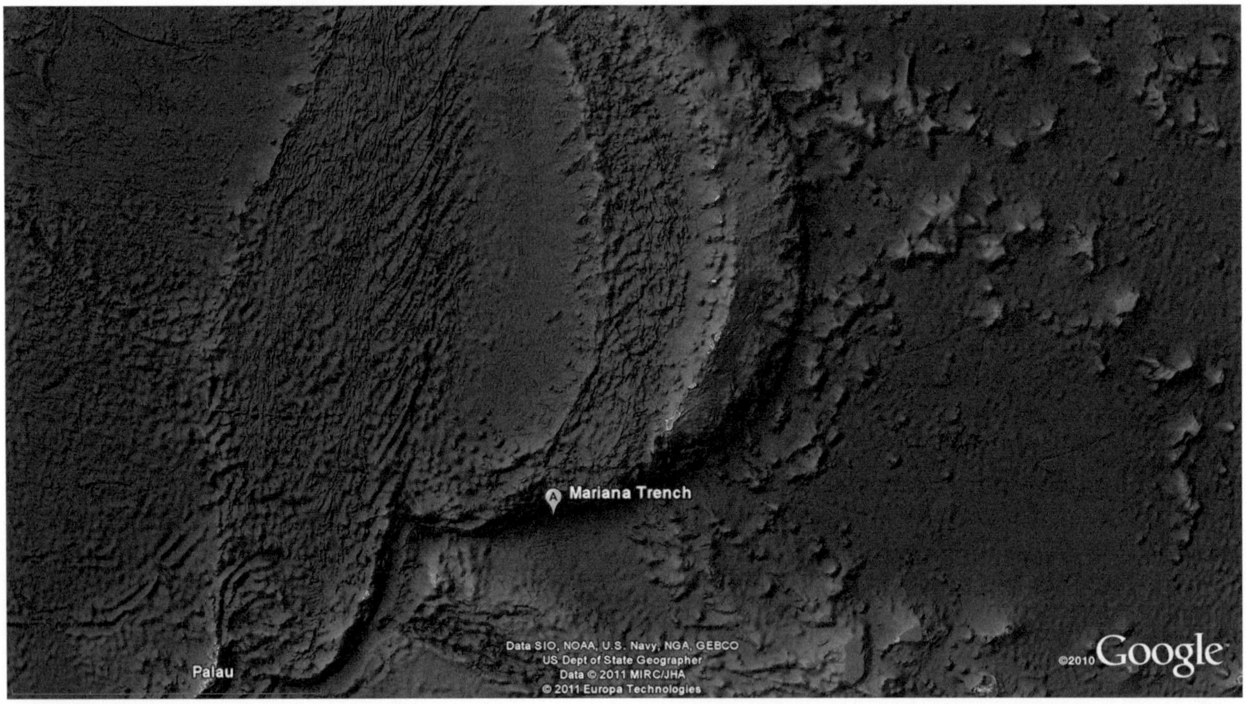

Fig. 1.11 The Mariana Trench is located in the Pacific Ocean where two oceanic plates converge and one descends beneath the other in a process known as subduction. The deepest part of the Mariana Trench is the Challenger Deep, so named after the exploratory vessel HMS Challenger II; a fishing boat converted into a sea lab by Swiss scientist Jacques Piccard. The Challenger Deep is the deepest part of the earth's oceans, and the deepest location of the earth itself. (Image credit: ©Google Earth 2010).

steep sides of the seamount). Therefore, plankton biomass is often high over seamounts which mean that they can attract large numbers of fish.

As the deep oceans are far from being sufficiently well understood, some of the processes are in a state of early exploration. Researchers were for example startled when they found natural hot springs in the sea (on land you may know natural hot springs such as Old Faithful at Yellowstone National Park). Similar phenomena occur under the oceans within mid-ocean ridge volcanoes where they are called deep-sea hydrothermal (hot water) vents or "black smokers" (Fig. 1.12). The latter are made up of sulphur-bearing minerals that have come from beneath Earth's crust. They form when hot (roughly 350°C) mineral-rich water flows out to the ocean floor at a mid ocean ridge volcano. Sulphide minerals grow or crystallize from the hot water and form a chimney- like sulphide structure through which the hot water continues to flow. As the hot, mineral-rich water mixes with the cold ocean bottom water, it precipitates a variety of minerals as tiny particles that make the vent water appear black in colour, hence the term black smokers. The water flowing out must exceed the high pressure of several thousand meters of water column, in which the bottom temperature is normally about 2°C. In spite of the extremely high temperatures, large tube worms, crabs, mussels and even fish live close to these vents feeding on chemo-autotrophic organisms. They represent a rare ecological niche separated entirely from the energy of sunlight.

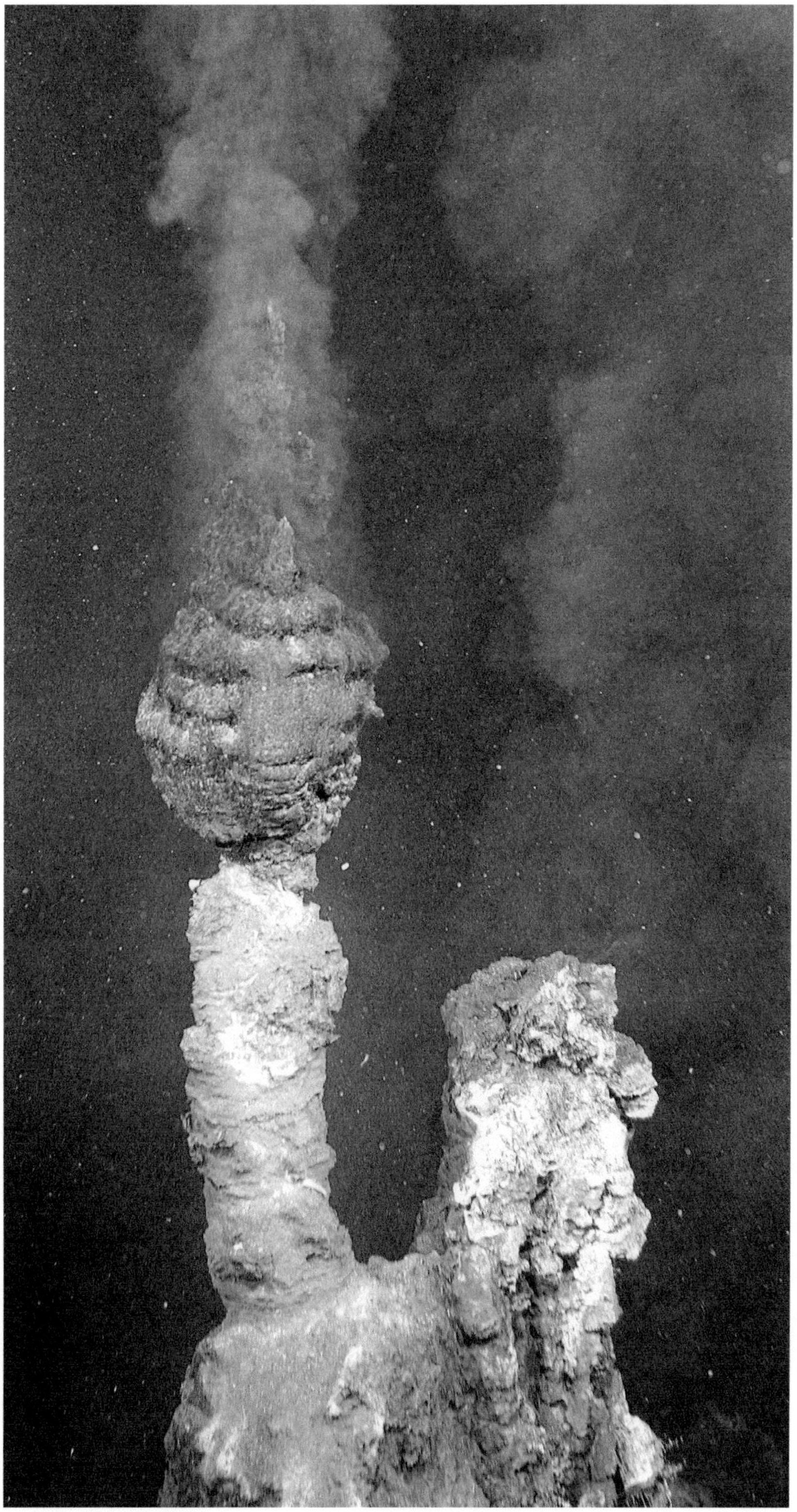

Fig. 1.12 In fault zones on the deep ocean floor "black smokers" send hot water up from the sea floor at more than 300°C as brines with dissolved metals. These chimney-like features are a habitat for a food chain starting with chemoautotrophic bacteria, and involves mussels, crabs, giant tubeworms and fish. (Photo credit: New Zealand American Submarine Ring of Fire 2007 Exploration, NOAA Vents Program, Institute of Geological & Nuclear Sciences and NOAA-OE).

1.2 Sediments in the Oceans

Over vast areas, the ocean floor is covered with thick sediments from different sources (Seibold & Berger, 1982). Lithic sediments that accumulate on the continental shelves and slopes consist of silts and clays that are delivered to the oceans by rivers from continental sources. Their load amounts to 18.3 billion tons/year. The focal points of sediment accumulation are river mouths and in particular deltas in tropical areas with a large discharge area from high mountains, such as the Ganges-Brahmaputra system or the Amazon basin. Other sources of lithic sediments are debris-laden glaciers and icebergs that raft sediments seawards of glacier margins (about 2 billion tons/year).

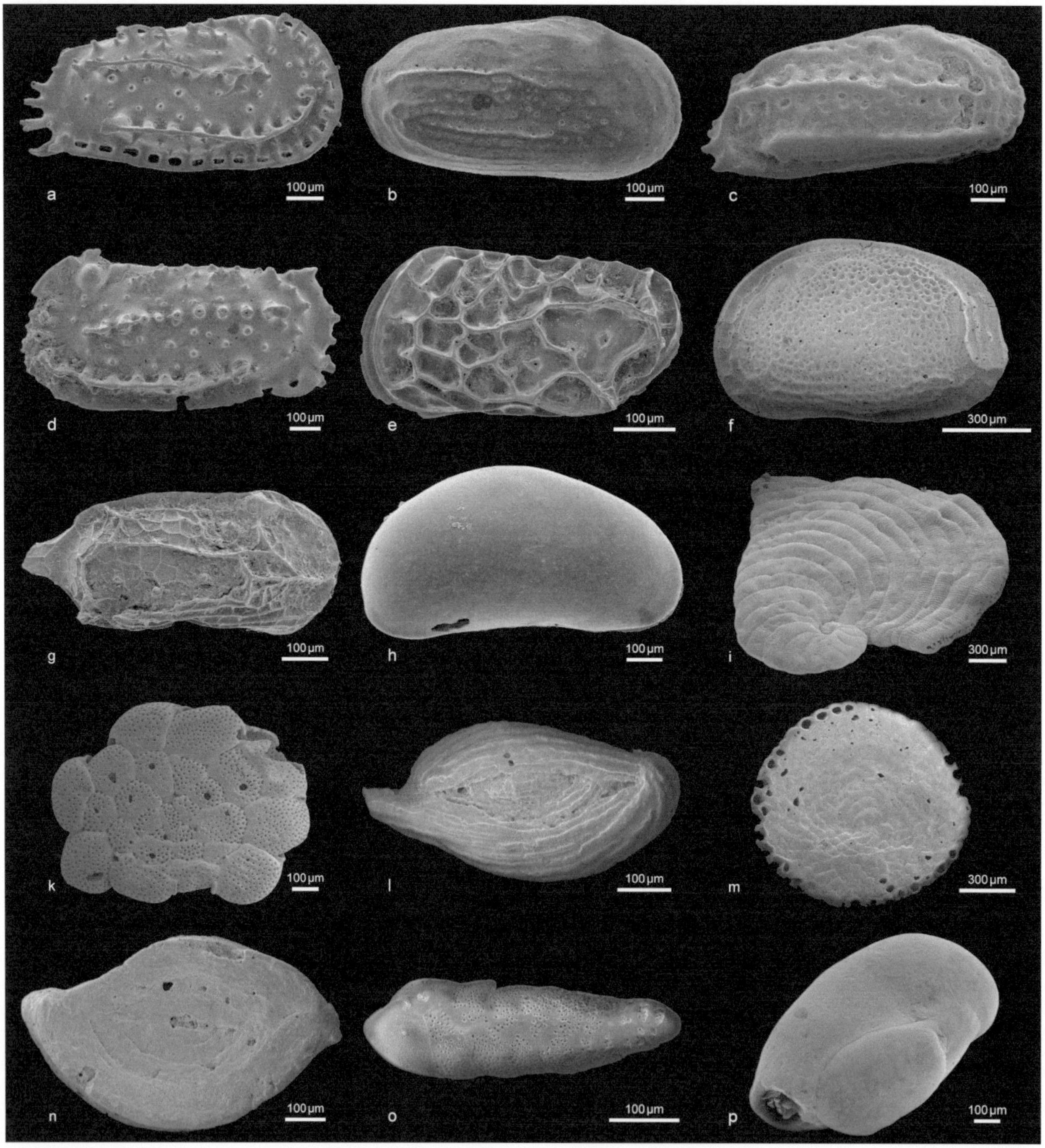

Fig. 1.13 Foraminifera made up of carbonate are mostly planktonic but may live on the sea floor. (Photo credit: Andreas Vött, University of Mainz, Germany).

Large areas of the deep sea floor are mantled with sediments that consist of the skeletal remains of tiny organisms such as single-celled planktonic animals. Warm surface waters in the low to middle latitudes favour the growth of carbonate secreting organisms like *foraminifera* (Fig. 1.13). Their tiny shells accumulate at an average rate of about 1–3 cm per thousand years at the ocean floor. Large volumes of carbonate crystals are also precipitated inside the intestines of marine fish and are then excreted at very high rates, releasing this lesser-known, non-skeletal carbonate into the marine environment. The source of the carbonate in the ocean is the product of solution of limestone by weathering processes. The calcareous oozes are not found where the ocean basins are very deep as cold, deep water (under higher pressure) dissolves more carbon dioxide. As a result, these deeper waters are more acidic than surface waters and they dissolve carbonate particles such as the shells of *foraminifera*. The depth at which this occurs is called Calcite Compensation Depth, or CCD, and in the Pacific is about 4000 to 5000 m, whereas in the Atlantic Ocean the CCD is somewhat shallower. Thus, all limestones that we now find in the terrestrial environment, even folded up into high mountains, were formed by the sedimentation of carbonates in rather shallow water. This is the case with corals at 3000 m above present sea level in the European Alps. In other regions, organisms that precipitate siliceous skeletons may dominate. These tiny *radiolarians* (animals) and *diatoms* (plants) (Fig. 1.14) are the major component of ocean sediments in regions with high biological productivity, as for example around the Southern Ocean or the equatorial Pacific and Indian Oceans. As sea water is undersaturated with respect to silicate, these shells would ultimately be dissolved, but this process is slow, and in nutrient rich coastal waters the bio-production is rapid, so that net accumulation normally occurs.

Far away from the continents and in regions of low productivity, the deep seafloor is covered with very fine grained clay, termed "red clay" as its colour is the result of oxidation of iron rich minerals in the sediment. The source of much of this clay is fine windblown dust that drifts over long distances with air masses from the continents over the ocean basins. Overall the sedimentation rate of these clays is very low with rates of only 1 mm to 1 cm over a million years!

Fig. 1.14 Diatoms are made of silica and are common as plankton. (Photo credit: Jan Michels, Institute of Zoology, Christian-Albrechts-Universität Kiel, Germany; Kathryn Taffs and Jo Green, Southern Cross University, Australia).

Fig. 1.15 Manganese nodules from a Tertiary deep ocean floor, now exposed on land in the Negev Desert of Israel. These nodules have a diameter of several centimetres and are composed primarily of manganese and iron. (Photo credit: D. Kelletat).

Another group of ocean sediments are derived from precipitation of dissolved minerals in water and are therefore called hydrogenous sediments. They may occur in all depths, but mostly as supersaturation by evaporation (as salt and gypsum) in shallow embayments or enclosed basins with a dry and hot climatic environment. On the deep ocean floors, over a very long time (millions of years), minerals may form nodules or concretions (Fig. 1.15) containing cobalt, manganese or chrome (Fig. 1.16). Efforts have been made to expand ocean mining into deep-sea waters. The focus has been on manganese nodules, which are usually located at depths below 4000 m, gas hydrates (located between 350 and 5000 m), and cobalt crusts along the flanks of undersea mountain ranges (between 1000 and 3000 m). However, the industrial countries have lost interest in these resources as prices for these minerals have dropped and new discoveries have been made on land with easier access.

1.3 Physics and Chemistry of Ocean Waters

As water is vital for life on Earth, the Blue Planet in our Solar System, a very short summary of its main physical and chemical properties will be given here. Water is a unique molecule and behaves differently from most other chemical compounds. The a-symmetrical atomic arrangement in a molecule of pure water (H_2O) is responsible for its special physical characteristics and a critical factor for climate. Its oxygen atom (O) and the two hydrogen atoms (H) have an angle of 104.5°. This produces a dipole, a molecule with one negatively and one positively charged end. Because the water dipoles tend to hold together like small magnets, water reacts sluggishly to warming or cooling. For a given amount of heat absorbed, water has a lower rise in temperature than other substances; it has, in other words, a high heat capacity. In fact, water has the highest heat capacity of all liquid and solid substances with the exception of ammonia (World Ocean Review, 2010). That is why water can absorb and release large amounts of heat with very little change in temperature, thus causing the inertia of climate processes influenced by the gigantic heat reservoir of the ocean. But only 37% of the sun's radiation penetrates the water column to 1 m depth, 16% to 10 m depth and only 0.5% to 100 m depth (if the water is very clear) (Fig. 1.17). Global sea surface temperatures, show pronounced east-west temperature belts approximately paralleling the equator. During August, the warmest waters exceed 28°C and occur in a belt between 30°N and 10°S latitude where the solar radiation is at its maximum. During winter in the Northern Hemisphere, this zone shifts south together with the belt of warm water until it is largely below the equator.

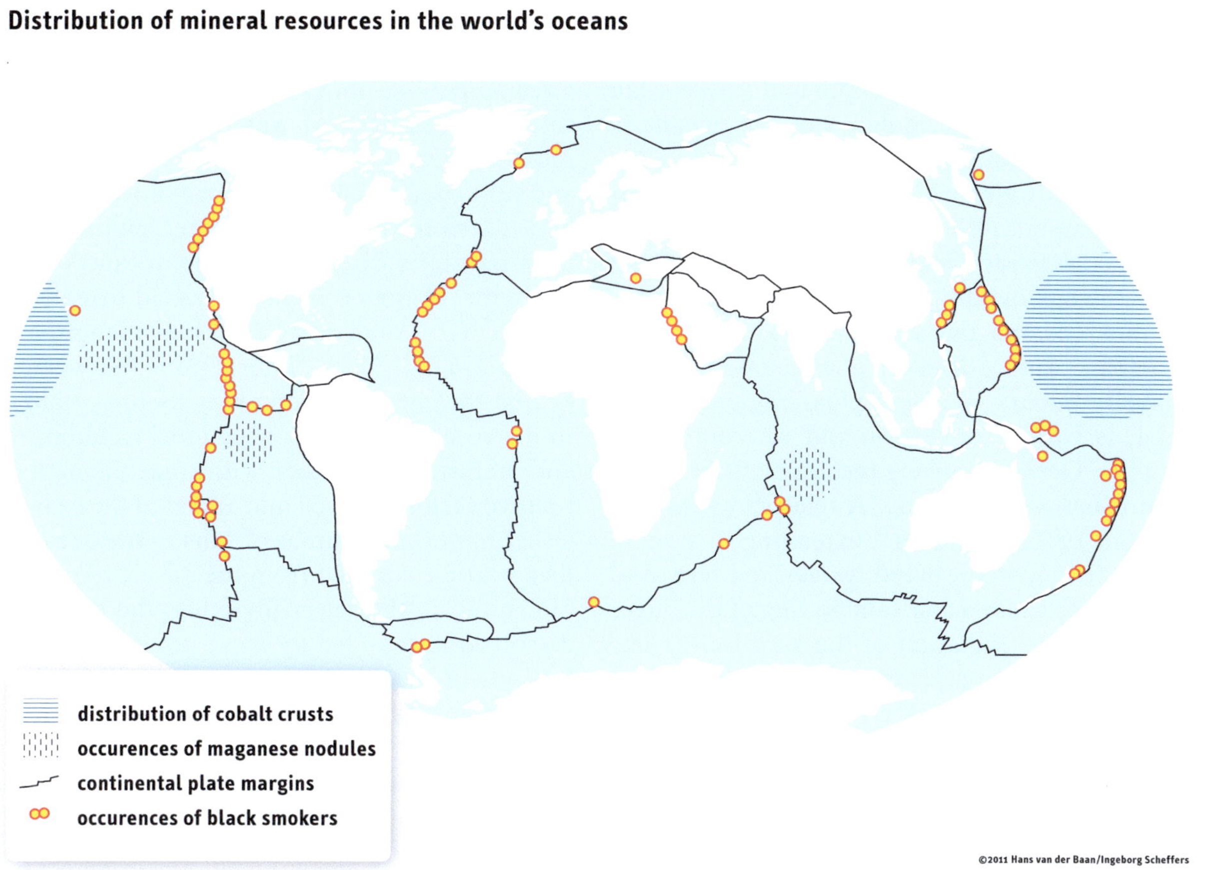

Fig. 1.16 Distribution of mineral resources in the world's oceans.

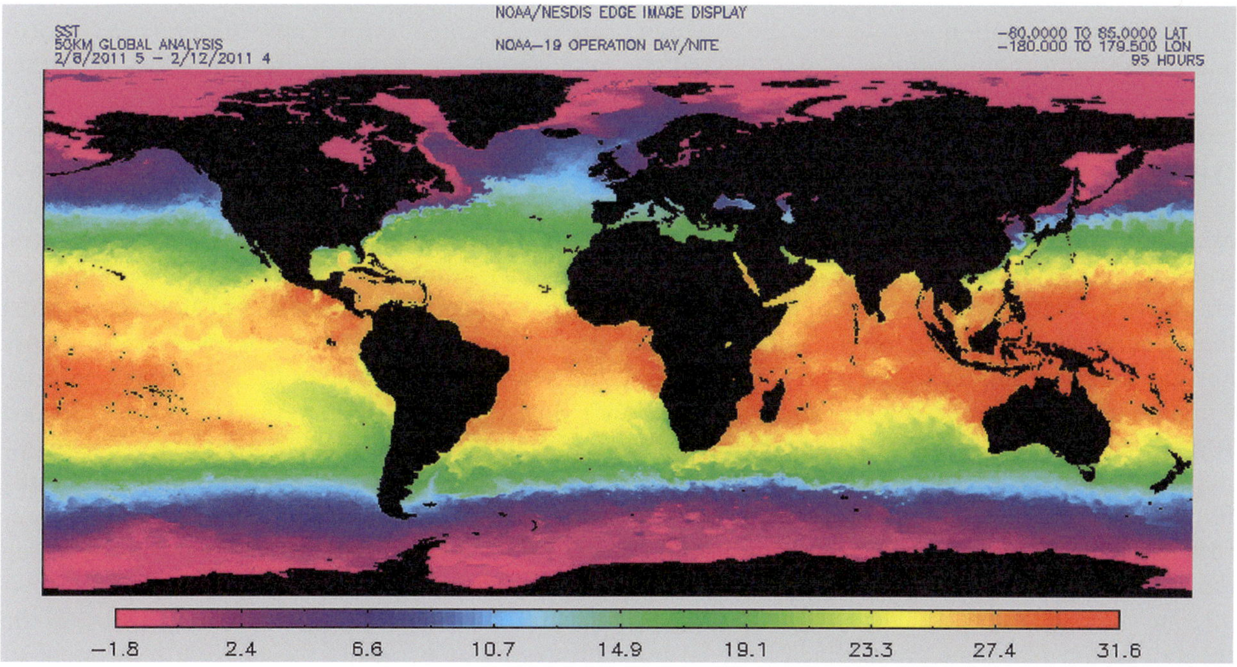

Fig. 1.17 Global sea surface temperatures in August. This sea surface temperature map was produced using MODIS data acquired daily over the whole globe. The red pixels show warmer surface temperatures, whereas yellows and greens are intermediate values, and blue pixels indicate cold water. (Photo credit: NASA's Visible Earth, http://visibleearth.nasa.gov/).

Oceans play a central role in the climate system. Beside their heat capacity, the physical properties of seawater vary with depth and drive ocean circulation. The water column of the oceans is vertically layered as a result of variations in the density of seawater which is a function of salinity and temperature. Cold, salty water is heavy and sinks to great depths until it reaches a level where the surrounding water has the same density thus causing the stratification of water in the ocean. This powerful phenomenon, which primarily occurs in a few polar regions of the ocean, is called convection and we come back to it later. Oceanographers recognize three major depth zones in the oceans: A relative warm surface zone or "mixed layer", extending to a depth of 100–500 m, where wind, waves and temperature changes cause a constant mixing of this layer warmed by the radiation of the sun. Below lies a zone in which temperature, salinity and density undergo significant changes with increasing depth. This zone is called the thermocline, a zone in which temperature decreases with depth. Under this lies the deep zone, which contains about 80% of the ocean's water volume.

The most important chemical constituent of ocean water is salt with a mean concentration of 35‰ (varying between 33 and 37‰). However, salinity is closely related to latitude and is controlled by: precipitation as rain and snow which adds freshwater making the seawater less salty; evaporation, which removes freshwater and makes the ocean water more salty; inflow of rivers carrying freshwater into the sea; and the freezing of sea ice because during the salt minerals are excluded from the ice leaving the unfrozen seawater more salty. Even if the salt content in general differs, the chemical substances are stable in their percentage: chlorides of sodium and magnesium comprise 88.7% of the total salt content, sulphates (of magnesium, calcium and potassium) 10.8%, and carbonates and bromides the remaining 0.5%. Nearly all chemical elements can be found dissolved in ocean water, including gases like oxygen and nitrogen and rare metals. If all the salt were precipitated, it would form a layer of 53 m over the entire seafloor. But where does the salt in the sea come from? Each year rivers and streams carry an estimated 2.5 billion tons of dissolved substances to the sea. This includes cations such as sodium and potassium which are leached out by weathering processes of rocks and become part of the dissolved load of rivers and streams flowing into the sea. Volcanic eruptions release gases such as water vapour and carbon dioxide, but also two important anions, chloride and sulphate, which dissolve in atmospheric water and return to the surface as precipitation, much of which falls directly into the ocean. Volcanic gas with its anions is also released directly into the ocean by submarine eruptions along the mid ocean ridges. Here, interactions between the heated rock and sea water play an important role in the composition of sea water as calcium, iron, and manganese together with trace elements are removed from the rock and added to the seawater. Other important sources of salts in the oceans are dust particles eroded from the desert regions and blown out to the sea. Altogether, the quantity of dissolved ions added by these processes over the billions of years of Earth's history exceeds the amount now dissolved in our today's oceans. But over time the composition of seawater remains virtually unchanged! The reason is that the substances are removed at the same time as they are added: Aquatic plants and animals are withdrawing elements such as silicon or calcium to build their skeletons; other elements like potassium or sodium are absorbed or removed by clay particles as they settle slowly on the sea floor, still others are precipitated to form new minerals in the oceanic sediments. Overall, the processes of extraction are equal to the combined inputs and the salinity of the sea remains unchanged over time. An open question is if the oceans always have been salty? Best evidence of past saltiness are evaporates of marine origin, which are common in young sedimentary basins, but are not present in rocks older than about one billion years. Most geologists agree that there is evidence in many ancient strata that evaporates once were present.

Salt concentration in the oceans has also a major influence on marine organisms. The maintenance of cell fluid composition is crucial to survival of marine organisms (as it is to freshwater life). Their body cells must have a means by which to adapt to changing salt concentrations in their environments. This balance is met through the processes of osmosis, the passive movement of water particles from a region of higher concentration to a lower concentration across a semi permeable membrane. Osmose regulation is the active regulation of particles within a cell enabling the organism to maintain sufficient osmotic pressure on their cell walls.

1.4 Life in the Oceans

The world's oceans are by far the most extended living space (about 99% of all Earth, Fig. 1.18) and are home to a huge variety of life forms. During the last ten years one of the most exciting international research programs, the "*Census of Marine Life, A Decade of Discovery*", in which 2,700 scientists participated, produced the most comprehensive inventory of life forms in the world's oceans ever compiled and catalogued as a basis for future research – all in all 28 million records and counting! More than 80 nations participated with 540 expeditions from pole to pole and in all environments of the oceans. The results have been communicated with 2,600 scientific publications. Learn more about this project on the Internet at http://www.coml.org/about-census and click on the Layer Census of Marine Life in the Gallery Layer of Google Earth.

This first baseline picture of ocean life can be used to forecast, measure, and understand changes in the global marine environment, as well as to inform the management and conservation of marine resources. The Census investigated life in the global ocean from microbes to whales, from top to bottom, from pole to pole, bringing together the world's pre-eminent marine biologists, who shared ideas, data, and results. During their ten years of investigation, Census scientists discovered new species, habitats, and connections and unlocked many of the ocean's long-held secrets. They found and formally described more than 1200 new marine species, with another 5000 or more in the pipeline awaiting formal description. They discovered areas in the ocean where animals congregate, from white shark cafés in the open ocean to an evening rush hour in the Mid-Atlantic Ridge and a shoal of fish the size of Manhattan off the coast of New Jersey, USA. They unearthed a biosphere in the microbial world, where rare species lie in wait to become dominant if change goes their way. While unlocking many secrets, investigators also documented long-term and widespread decline in marine life as well as resilience of the ocean in other areas where recovery was apparent.

Warm and cold water ocean currents

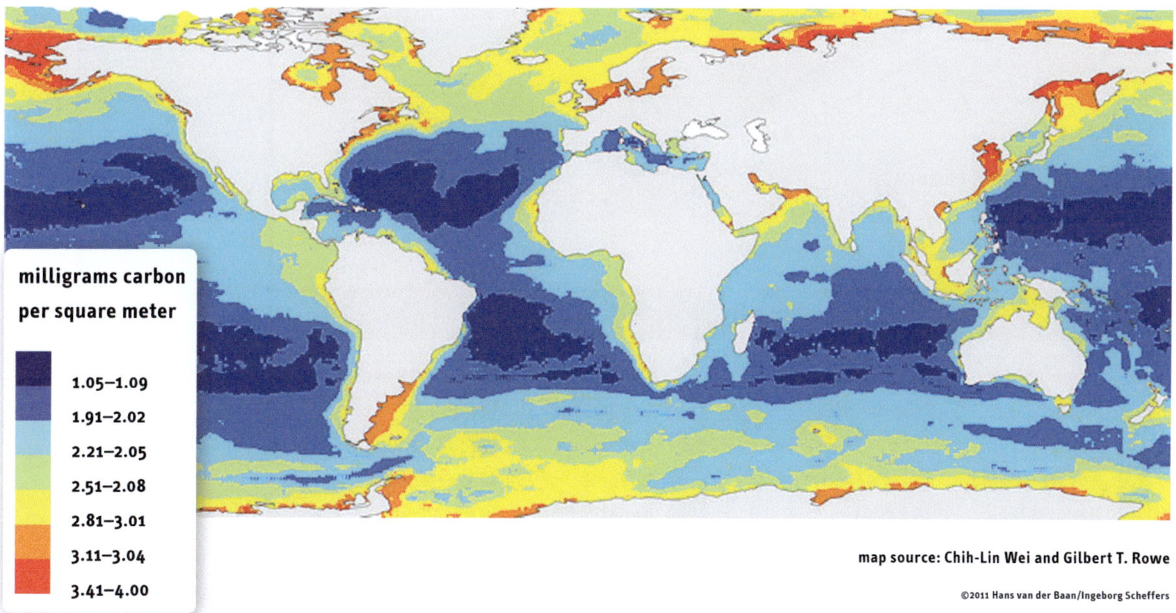

Fig. 1.18 World map of life in the oceans depicting biomass – from bacteria to fish. Biomass clearly dominates in coastal areas because of elevated nutrient and sun light levels in shallow waters, as well as in cold regions due to increased dissolved oxygen. (Source: AFP, Census of Marine Life: Chih-Lin Wei, Gilbert T. Rowe).

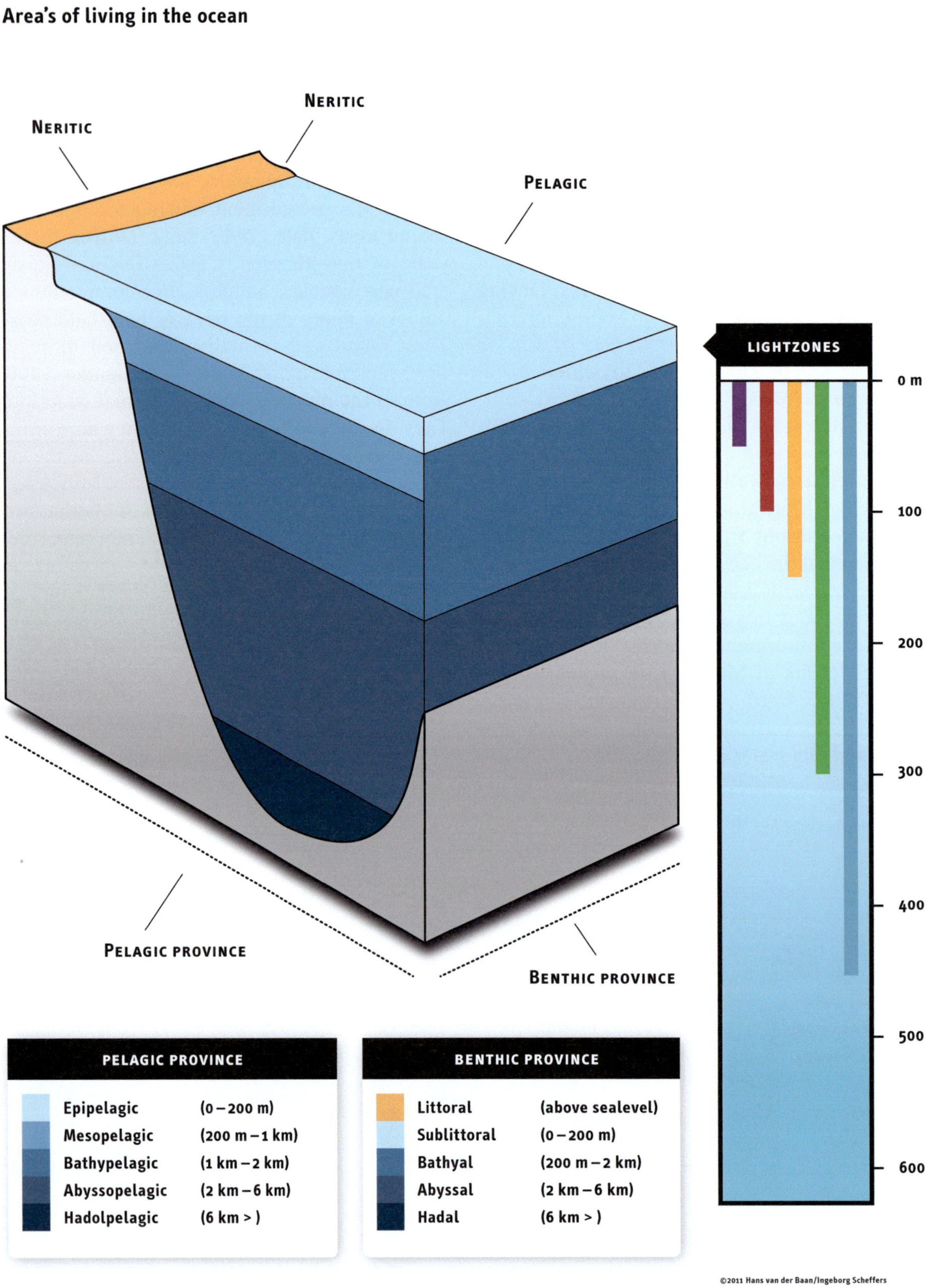

Fig. 1.19 The most significant ecological regions of the oceans in terms of bioproductivity are the littoral and neritic zones (shorelines and shelves), the benthic zone (the sea floor in all depths), and the uppermost layer of ocean water with high oxygen, nutrient and sunlight levels (the photic or euphotic zone).

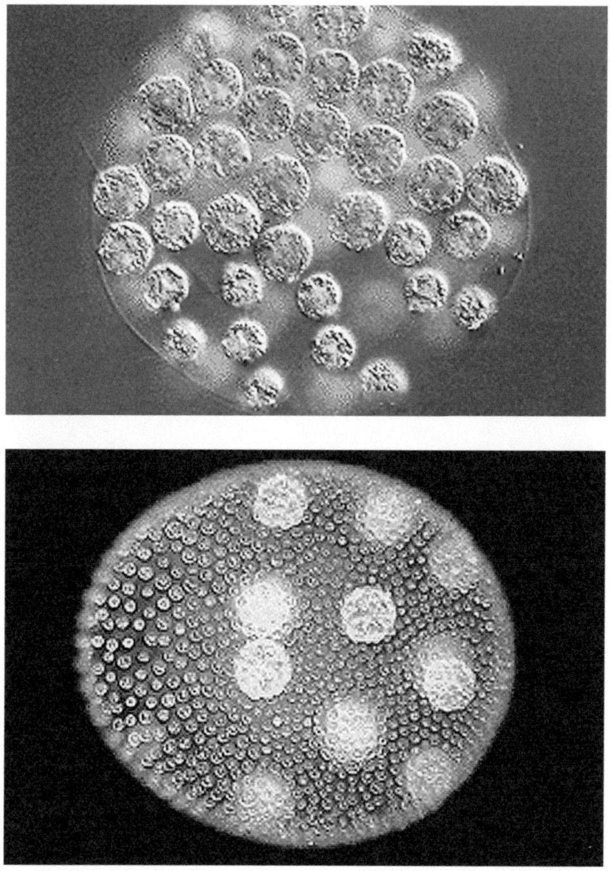

Fig. 1.20 Green algae contain chlorophyll and produce oxygen, using sunlight. (Photo credit: David Kirk, www.ncbi.nlm.nih.gov).

Nearly all of the ocean surface waters carry microscopic plants as phytoplankton (Figs. 1.19 and 1.20). Phytoplankton converts nutrients into plant material by using sunlight with the help of the green pigment chlorophyll. This microscopic plant life is at the base of the marine food web and is the primary food and energy source for the ocean ecosystem. Around half of the worldwide primary productivity is achieved by these microscopically small plants, which grow and multiply in the ocean. Small fish and other animals eat them as food, larger animals then eat these smaller ones. The ocean fishing industry often finds good fishing spots by looking at ocean color images which can reveal phytoplankton blooms.

Marine animals show high species diversity and abundance in the oceans, and – like the plants – many of the species are yet to be described. The animals of the sea are divided into three groups: those that only drift (or marginally support themselves by flagella) such as *microzooplankton* (e.g. foraminifera); and other so-called *meso-* and *macrozooplankton* (e.g. copepods and jellyfish respectively) that are less dependent on water movement. Those that actively swim and are independent of water movement are called *nekton* (e.g. fish and marine mammals like seals or whales), and those that live on or in the bottom sediments are called *benthic organisms* (e.g. sea urchins, gastropods, lobsters, worms, molluscs and many others). Some of these are sessile and unable to move (barnacles, oysters, corals), others such as molluscs or sea stars may move but nevertheless depend on the substrate and nutrients that only can be found at the sea floor.

The open water biotic zone is also termed the *pelagic* zone of the ocean. This can be thought of in terms of an imaginary water column that goes from the surface of the sea almost to the bottom, as shown in Fig. 1.19. Plants and animals living in this zone are called pelagic organisms. This pelagic environment comprises far more than 1 billion km³ and is divided into different zones, depending on depth and therefore the amount of light available. In general, the first metre is called the *euphotic* zone which is the most productive layer in the oceans but its depth can vary tremendously, depending on a number of different factors. More than half of the sunlight is absorbed in this zone and red light, which all green photosynthesizers such as green algae use, is all absorbed here. Plant life is restricted to the upper 100–200 m *disphotic* zone of the ocean. Here sufficient light (energy) is available for the process of photosynthesis. After ten metres, 50% of the sunlight is absorbed and at 70 metres depth there is just enough light available for photosynthesis to keep the organisms alive, but there is no surplus energy left for repair, growth or reproduction. At 100–200 m depth only 0.5% of sunlight penetrates and photosynthesizers cannot live here. The *euphotic* and *disphotic* zones make up the *epipelagic* (sunlit) zone. The twilight zone down to about -1000 m is called *mesopelagic*. Here insufficient light restricts the process of photosynthesis and therefore this zone is depleted of oxygen. But organisms compensate for this by different adaptations such as for example, by slow movements. The bathypelagic zone extends down to -4000 m, where, with no sunlight or living plants, most of the organisms live from the detritus settling down from the higher zones. In the deepest sections, the *abyssopelagic* (midnight) and the *hadopelagic* (lower midnight) environments down to 11000 m depth, only very few organisms live

Fig. 1.21 Deep-sea coral habitats in the Aleutian Islands of Alaska. (Photo credit: Bob Stone, NOAA Fisheries/Marine Photobank).

in the free water column (as far as we know!). At the deep sea bottom near the hydrothermal vents life relies on bacteria that provide energy derived from the escaping sulphurous gases. Even in the darkness these ecosystems of the deep ocean can be vivid in colour and their uniqueness inspires advocacy to protect and preserve their fragile living environment (Fig. 1.21).

1.5 Movements in the Ocean: Currents, Waves and Tides

How does water move in the oceans? As you have seen, in the depth of our oceans water movement is driven by the characteristics of the water itself – its temperature and salinity. The forces that drive surface ocean circulation are fundamentally different: these currents are broad, slow drifts of surface water set in motion by the prevailing surface winds which also generate the waves on the water surface. They usually do not exceed depths of more than 50 to 100 m. The ultimate source of this motion is the sun, which is heating the surface of our planet unequally and thereby setting in motion the planetary wind system. The oceans are an immense heat reservoir that retains energy from the sun over a long time and the large ocean currents transport this heat for thousands of kilometres and significantly influence the climate in many regions of the world.

Before we discuss the current systems in more depth, we briefly touch on two influences on the direction of these water movements: The *Coriolis Effect* and *Ekman Transport*:

The Coriolis Effect results from the earth's rotation and causes all moving bodies (of water, air, or even cannon balls) to be deflected to the right in the Northern Hemisphere and to the left in the Southern Hemisphere. As the magnitude of the Coriolis force varies with latitude, the effect reaches a maximum at the poles and a minimum at the equator. Water masses that flow from the equator towards the North Pole in the Northern Hemisphere will be deflected toward the east (to the right) by this effect and water moving from the North Pole toward the equator will be deflected to the west (still to the right). The opposite happens on the Southern Hemisphere. The deviation of the Coriolis effect leads to quasi circles of surface currents with a right turn in the Northern Hemisphere and a left turn in the Southern Hemisphere. By friction along the sides of a current, meanderings and eddies occur (Fig. 1.22).

The "Ekman spiral" was first proposed (conceptually) by the great Norwegian explorer Fridtjof Nansen. During his polar expedition in the late 1890s, Nansen froze his ship *Fram* into the ice north of Spitsbergen Island and allowed it to drift for more than two years. During the expedition he noticed that the drift of the ice was generally to the right of the wind. The Ekman spiral is a theoretical model of the effect on water of wind blowing over the ocean (Fig. 1.23). As we just discussed due to the Coriolis effect the surface ocean layer is expected to drift to the right in the Northern Hemisphere. Water in the lower layers drifts as well, however not as fast as the surface water because internal friction cause a decrease in the effectiveness of wind with depth. As a result the Coriolis effect becomes relatively stronger and stronger, deflecting each, slower moving layer farther to the right. In an ideal case, a steady wind blowing across an ocean of unlimited depth and extent causes surface waters to move at an angle of 45 degrees to the right of the wind in the Northern Hemisphere (45 degrees to the left in the Southern Hemisphere). At a depth of about 100 to 150 m water moves so slowly (about 4% of the surface current) that it is deflected in a direction opposite

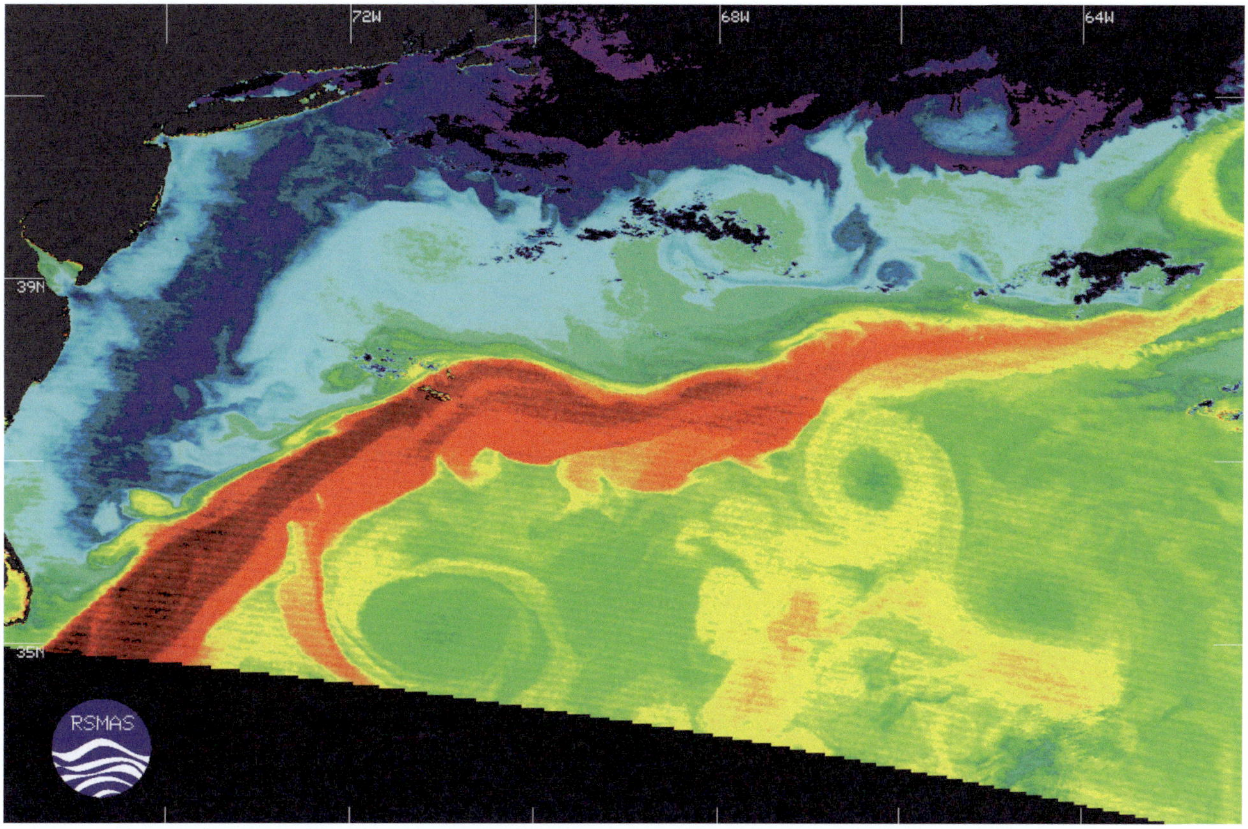

Fig. 1.22 Satellite images show that ocean currents can have complicated patterns. The depicted Gulf Stream is a strong ocean current that carries warm water from the sunny tropics to higher latitudes. This current extends from the Gulf of Mexico up the East Coast of the United States (Florida current) and departs from North America south of the Chesapeake Bay and heads across the Atlantic to the British Isles. The water within the Gulf Stream moves at a stately pace of four miles per hour. Even though the current cools as the water travels thousands of miles northward, it remains warm enough to moderate Northern European climate. The coldest waters are shown as a purple colour, with blue, green, yellow, and red representing progressively warmer water. Temperatures range from 7 to 22 degrees Celsius. (Photo credit: The sea surface temperature image was created at the University of Miami using the 11- and 12-micron bands, by Bob Evans, Peter Minnett, and co-workers for NASA's Visible Earth, http://visibleearth.nasa.gov/view_rec.php?id=215).

that of the wind. This is considered to be the lower limit of the wind's influence on ocean movement. The average flow over the full depth of the spiral is called Ekman transport and results in a net direction of water movement at about 90° to the wind direction.

1.5.1 Ocean Currents

Recall that the oceans are vertically stratified according to density (a function of salinity and temperature) and cold, salty water is denser than warmer and fresher water. The sinking of these cold, dense layers propels a global convection and circulation system. Convection connects two distinctly different components of the ocean: the near-surface layers that are in contact with the variable atmospheric fields of wind, radiation and precipitation, and the deep regions of the ocean. At the surface, currents, temperature and salinity fluctuate on a scale of weeks to months but at greater depths the environmental conditions change over time scales of decades or centuries.

The freezing of water in the Polar Regions plays a central role in driving the global circulation system. This is called the thermohaline circulation system because it is influenced by both the temperature and salinity characteristics of the ocean water (Fig. 1.24). When water freezes to ice, it only contains about five tenths of a per cent salt and this increases the salinity of the surrounding ocean water and thus its density. In the region around Iceland (in the Greenland and Labrador Seas) these cold waters of the Atlantic sink down as a kind of

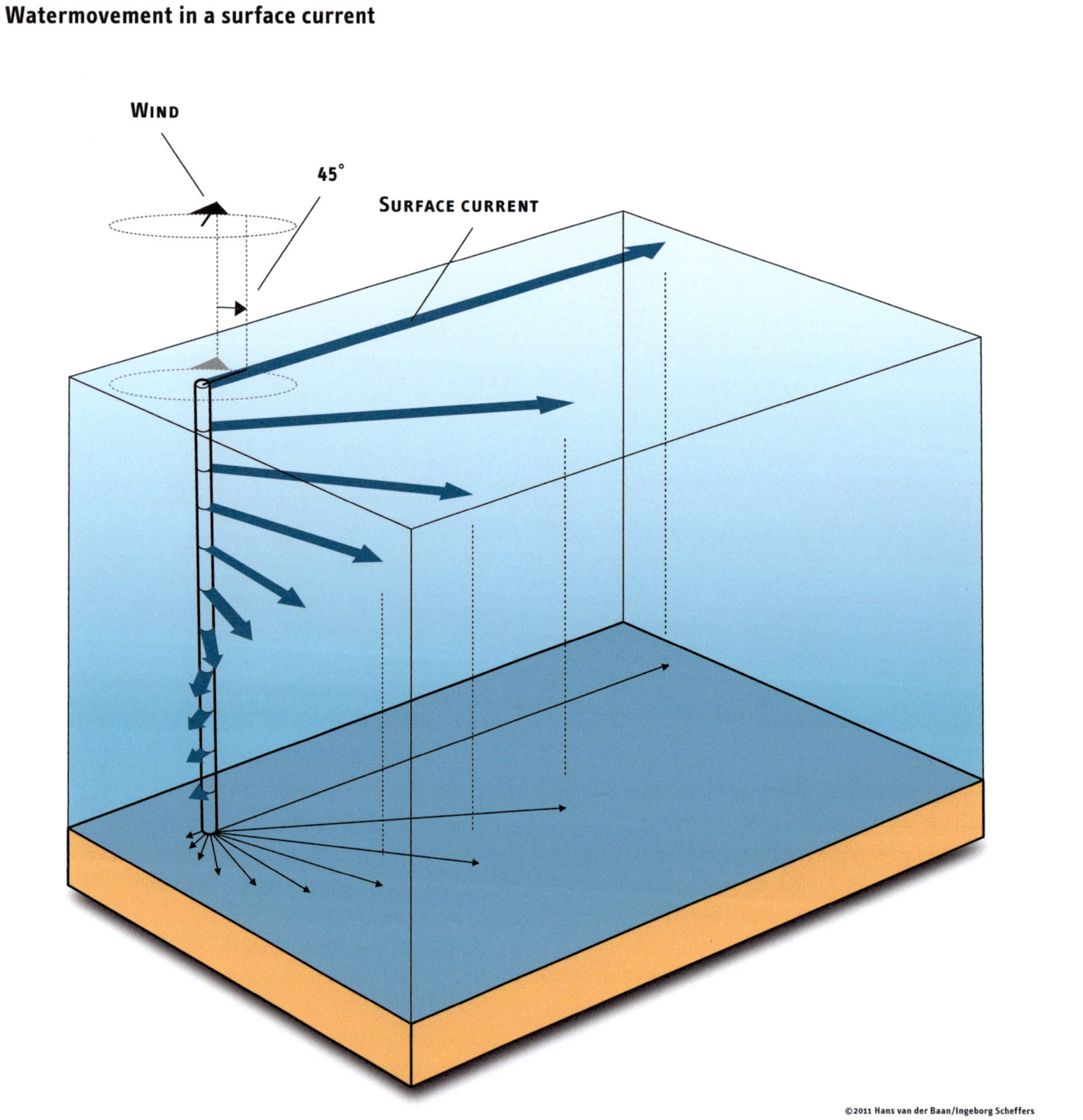

Fig. 1.23 The "Ekman Spiral", shows a complex system of water movements. Because of the Coriolis effect, a constant turning of the water mass (spiralling clock-wise in the Northern Hemisphere) can be observed.

elevator to the deep. This water mass produced by convection in the Arctic is called North Atlantic Deep Water (NADW). A similar convection cell exists in the Antarctic regions which produces the Antarctic Bottom Water (AABW) that flows across the ocean floor halfway around the globe into the North Atlantic. Because of their temperature and higher salinity the water masses produced here sink all the way to the sea floor and form the bottom layers, upon which flows the slightly warmer Arctic deep water. Both deep water currents move very slowly (at around one to three kilometres per day, due to their high density) along the abyssal plains towards the equator, where the temperature at 6000–7000 m depth is not higher than 2–3°C. In the near surface layers of the ocean a return flow of warm water occurs in the global conveyor belt of thermohaline circulation. The warm upper water from the east Pacific Ocean crosses north of Australia and through the Indian Ocean south of Africa and moves northwards to feed the Gulf Stream from Mexico to NW Europe. The amount of

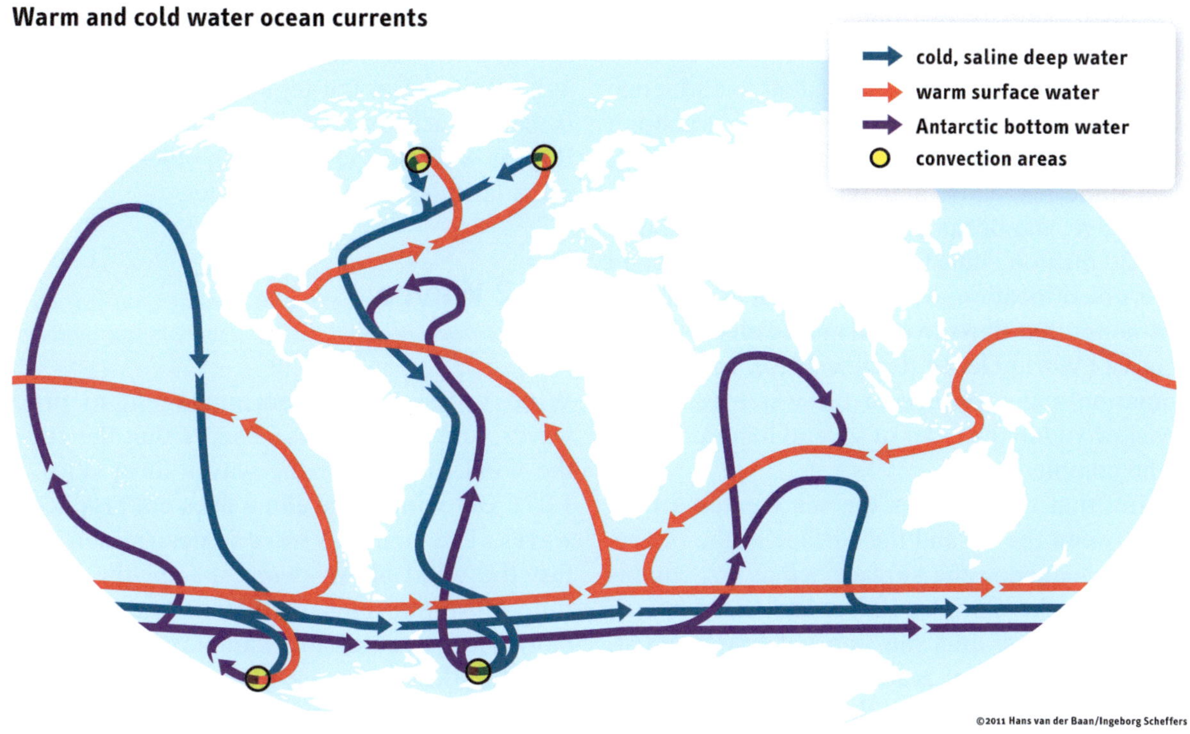

Fig. 1.24 All ocean currents are connected by the so called "conveyor belt", which is an important factor in driving Earth's climate.

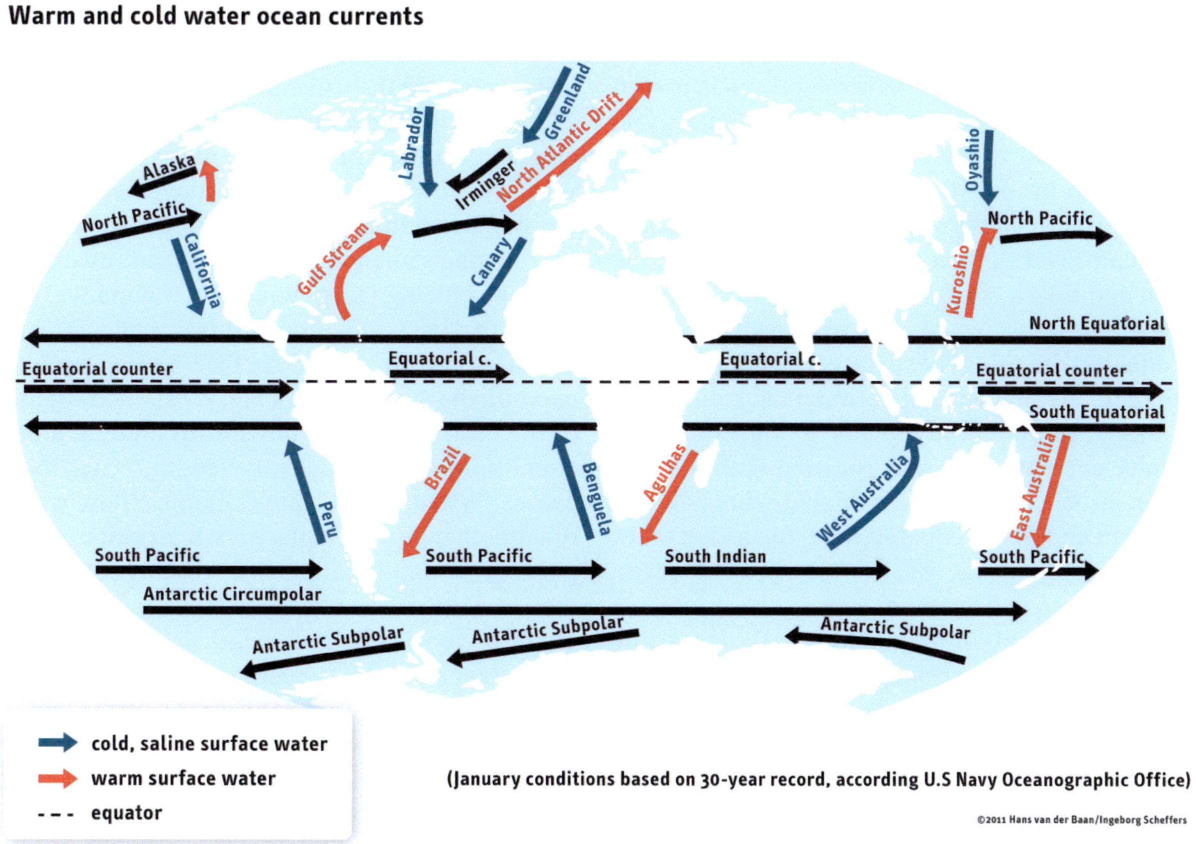

Fig. 1.25 A simplified picture of surface ocean currents.

water involved in the thermohaline circulation system is immense with a volume of around 400.000 cubic kilometres, which is equivalent to about one third of the total water in the ocean. This is enough water to fill a swimming pool 400 kilometres long, 100 kilometres wide, and ten kilometres deep. On average, the oceanic conveyor belt transports about 20 million cubic metres of water per second past a given location, which is almost 5000 times the amount that flows over Niagara Falls in North America (World Ocean Review 2010). In general, circulation and exchange of the water masses is very slow and in the order of several hundred years for the complete cycle.

Wind that flows across the sea creates a friction between the air and the surface of the ocean. As a consequence the surface winds, in combination with the Coriolis Effect and the shape of the ocean basins, drag the water slowly forward, creating a current of water as broad as the air current, but rarely more than 50–100 m deep. Consequently, the surface circulation pattern of the oceans is widely adapted to the global wind pattern (Fig. 1.25) and shows a distinctive pattern: These patterns curve to the right – clockwise – in the Northern Hemisphere and to the left – counter clockwise – in the Southern Hemisphere. Each major ocean current in both hemispheres is part of a large subcircular current system called a gyre. Five gyres exist in the world's oceans: two in the Pacific, two in the Atlantic and one in the Indian Ocean. Simplified we can say that on both sides of the equator warm, westward-flowing currents, the North and South equatorial currents, occur as the trade winds blow towards the west on either side of the equator, dragging the surface ocean water along with them. Sandwiched in between them and flowing eastward along the equator is the Equatorial Counter current which is associated with the doldrums, a belt of light and variable winds. When the North and South equatorial currents encounter landmasses along the western edge of the ocean basin they are deflected poleward. These western boundary currents flow parallel to the coastline towards the poles and transport enormous amounts of heat into the higher latitudes, significantly influencing the climate in many regions of the world.

In the Atlantic Basin, this current is called the Gulf Stream, a relatively fast current flowing along the coast of North America towards Europe. It reaches a speed of around 3.6 kilometres per hour at the sea surface, which is a casual walking speed! Europeans all benefit from the Gulf Stream as the climate in the region of the North Atlantic is comparatively mild, especially in northwest Europe whereas the winters in other regions at the same latitude are notably colder.

1.5.2 Waves

Wind is the most important agent to produce waves, and it is easy to observe that the stronger the wind, the higher the waves form (Figs. 1.26, 1.27). But wind force alone does not produce high waves. The size of a wave is determined by how fast the wind blows *(wind speed)*, the length of time *(duration)* and the distance across which it blows *(fetch)*. Think for example of a fetch of more than 1000 km , a strong wind (e.g. 12 Beaufort, or in other units 110–120 km/h) lasting for more than 36 hours – only then will waves with a maximum height of around a 20 m be developing in the open ocean, towering over ships unfortunate enough to be caught in them. Larger freak waves are possible, but very rare. The largest one so far has been measured at 34 m (vertical difference between the wave crest and the wave trough), with a velocity of 23 m/s (or 108 km/h) and a wave length (from one crest to the other) of several hundred meters. Wave velocity, however, is not the movement of water but the wave impulse propagation, which results in the phenomenon of "swell" far away from wave producing storm winds. In a wind wave, the water particles move on orbital tracks, of which the largest orbit at the surface has a circumference of the wavelength (Fig. 1.28). The diameter of the orbits decreases downward until, at a depth equal to about half the wavelength, water motion ceases. When a wave moves into shallow water, a drastic change takes place as the orbits of water particles at depth flatten and those in contact with the sea bottom simply move back and forth. As a result wave velocity and length decrease, the wave form steepens until it becomes unstable and spills against the shoreline as breakers or surf (Fig. 1.30).

A close relation exists between wave height, wave length and wave period (i.e. the time in which one wave crest follows the other at a given point): a wave height of 2 m has a length of about 50 m and a period of about 6 seconds, whereas waves 10 m high may have crests 200 m apart and a period of 12.5 seconds.

The global distribution of wave height and energy in the oceans is related to different climate systems (Fig. 1.31, 1.32): in calm regions or small water bodies and bays waves usually are small, whereas in the storm belts of the higher latitudes and in the cyclone belts of the lower latitudes they may reach considerable heights (Figs. 1.34, 1.35). Large wave systems run out of their wind generated areas over very long distances. These waves, occurring even in calm regions, are called swell and they are typical for many subtropical and

Fig. 1.26 Large and irregular waves in the open ocean during storm conditions. (Photo credit: NOAA/NSW).

Fig. 1.27 A 12 m-wave approaching a 5 m high cliff on Bonaire (southern Caribbean) during hurricane Ivan in 2004 (Photo credit: A. Scheffers).

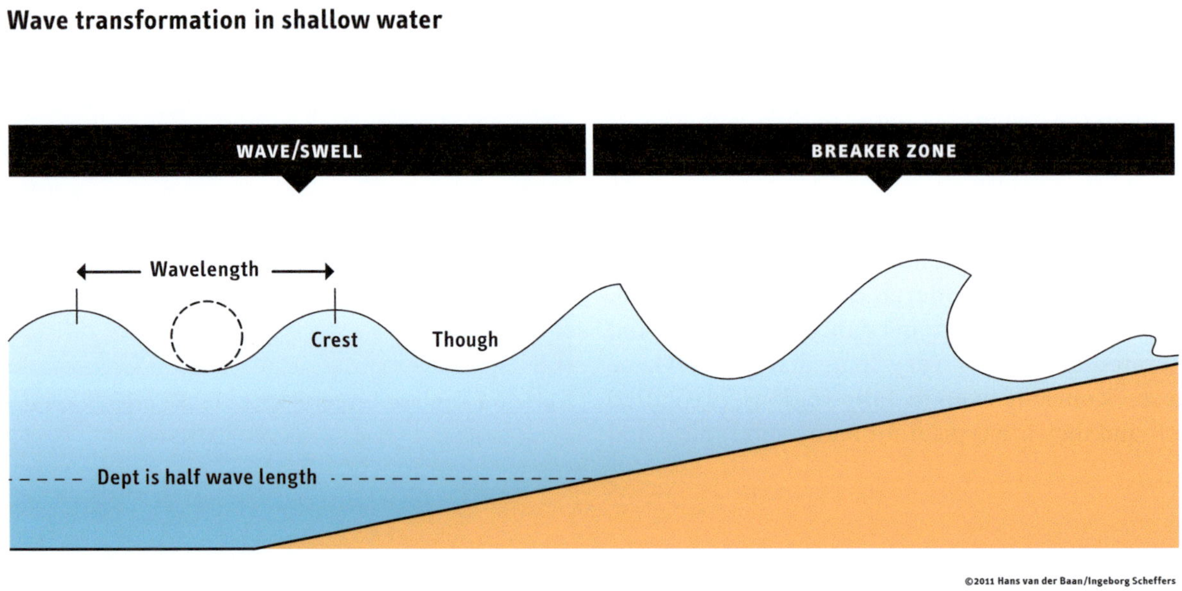

Fig. 1.28 These regular parallel wave patterns are called "swell".

Fig. 1.29 Strong surf at the south coast of Western Australia. The normally undulating forms are transferred into a typical surf belt in shallow water. (Photo credit: D. Kelletat).

Fig. 1.30a,b,c Orbital tracks of a high wave approaching shallow water are altered by friction at the bottom, whereas the upper parts of the orbital tracks move water forward into a plunging breaker. Kalbarri, Western Australia. (Photo credit: S. Scheffers).

tropical regions where they are deviated towards the equator by the Coriolis Effect from the higher latitude storm belts. The constant rhythm of these swell waves in warm latitudes are favoured by surfers as in the Bay of Biscay, Hawaii, or Tahiti.

When prevailing waves enter the shoreline at an oblique angle to the bathymetric contour, its crest bends to align with those contours, in a process called wave refraction (Figs. 1.35a,b,c). The part of the wave crest in the shallowest water is slowed the most, whereas the part of the wave in deeper water moves forward at higher velocity. Thus, the crest of the wave is bending towards the shore and the wave energy concentrates or dissipates at the shoreline. In general, wave energy is concentrated around headlands and spreads out while entering the bay at the beach over wide areas. If the waves break at an angle to the beach, a longshore current develops. The current is like a river on land and capable of moving sand along the beach through longshore sediment transport or also called littoral drift. Because the sediment grains are subject to both wave run up and littoral drift, they follow a zigzag path along the beach (Fig. 1.36). When an obstruction such as a jetty or groin is placed in the path of the longshore current, accretion of littoral drift occurs on the upstream side and erosion results on the downstream side. The extent of this accretion and erosion depends on the velocity and persistence of the current as well as on the supply of sand.

A different type of wave, so called shallow water waves (i.e. those with a large wave length compared to water depth) are triggered by sudden impacts on the ocean water column like earthquakes, submarine slides, volcanic collapse or meteorite impacts (summarized under the term of "tsunami" and discussed in Chapter 8).

Wave height at the coasts of the world

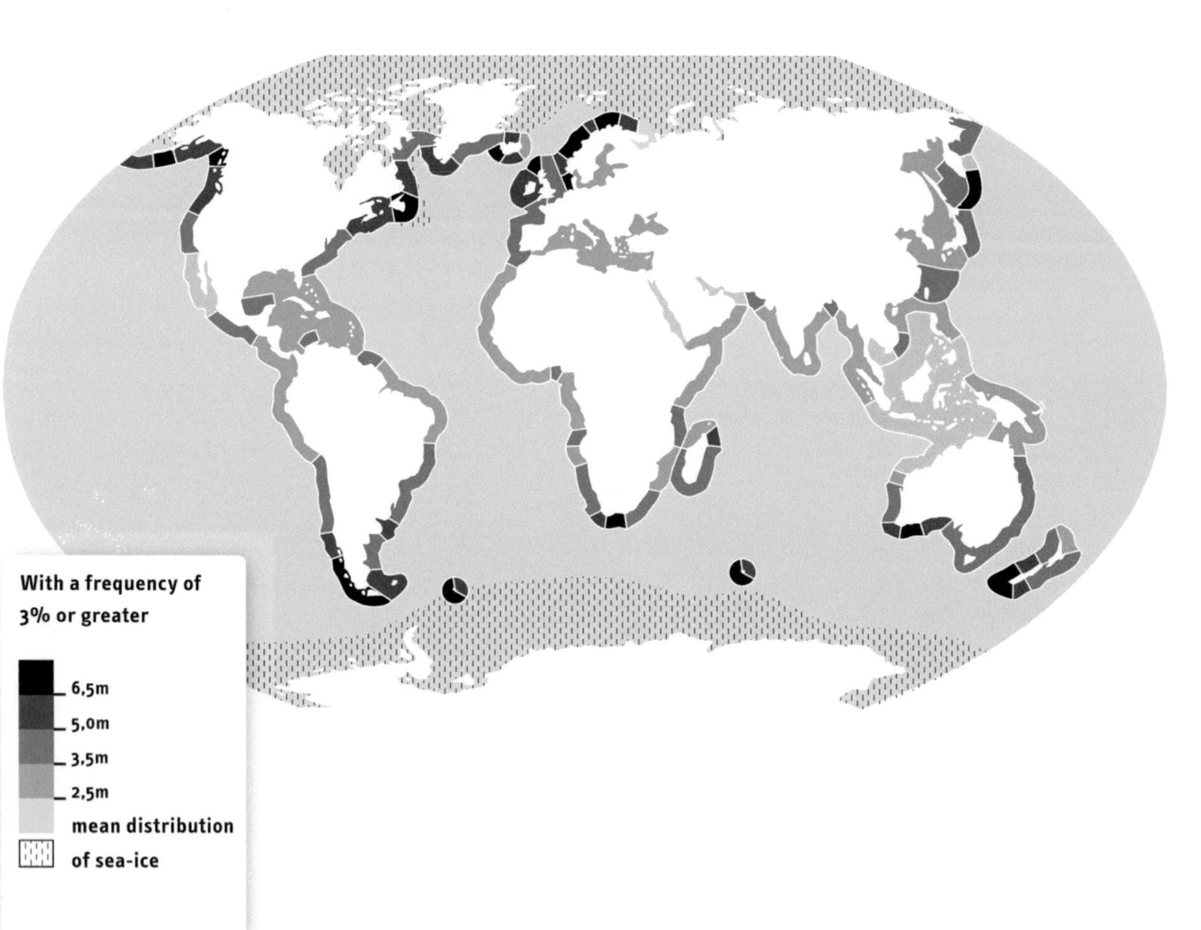

Fig. 1.31 Average wave heights at the coastlines of the world (Modified from Davies, 1980).

Global wave power distribution

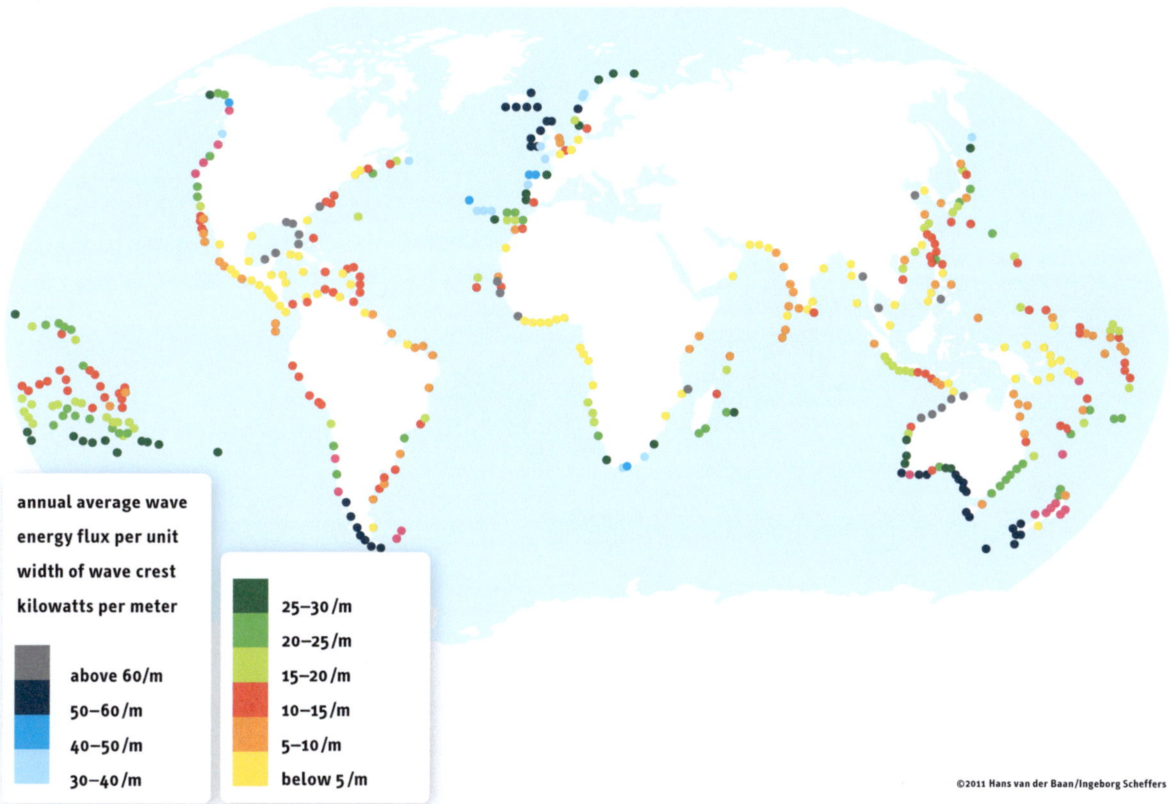

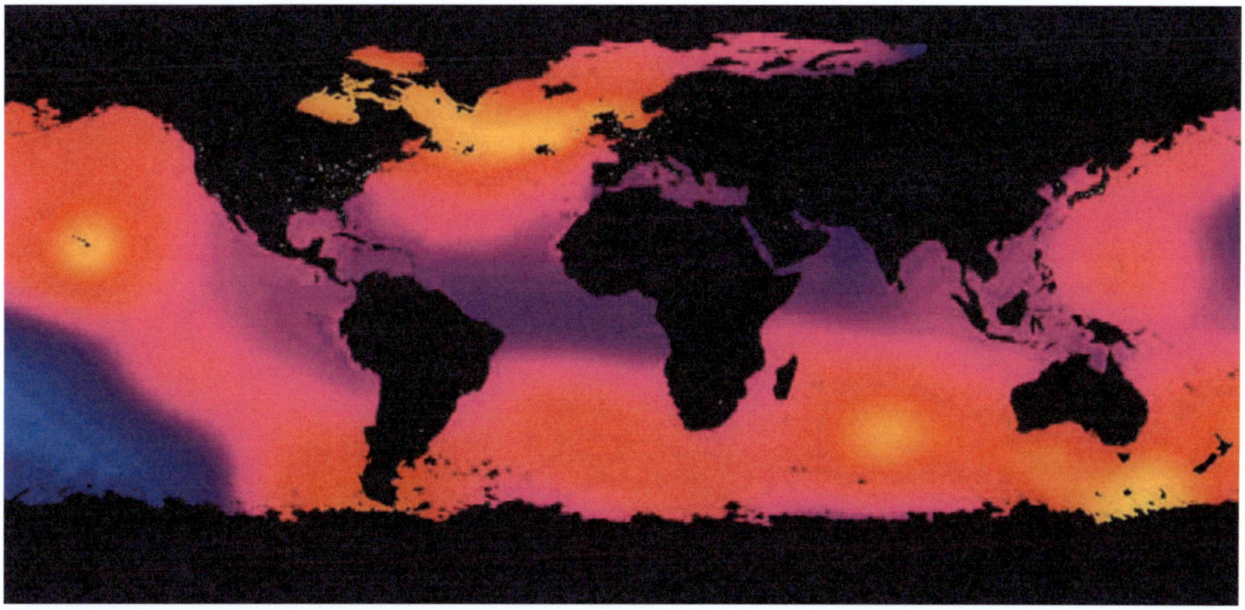

Fig. 1.32a,b Wave force is correlated to climate zones with high wind velocities. Wave energy is greatest towards the poles (yellow) and least at the equator (blue). This map is only a generalization based on data from the World Energy Council. It has been superimposed on a satellite composite of the earth at night (NASA) that shows where the current electricity demand is concentrated. (Image Credit: Benjamin Gatti, http://www.windwavesandsun.com/WaveResource.html).

Fig. 1.33a,b Shapes of breakers and water fountains during a strong hurricane at the 5 m high cliff of Bonaire's east coast, southern Caribbean, and backwash into the sea. (Photo credit: S. Scheffers).

Fig. 1.34a,b Water fountains from 4–5m waves during a summer storm at the Aran Islands of western Ireland, rising to a cliff top at +50m asl. (Photo credit: D. Kelletat/A. Scheffers).

NEXT PAGE

►Fig. 1.35a,b,c Wave refraction by obstacles (promontories islands). (Image credit: ©Google Earth 2010).

►Fig. 1.36 Swell from the NW at the coast of Venezuela produces longshore drift to the south. The orientation of the wave crests perpendicular to the shore can be clearly seen. (Image credit: ©Google Earth 2010).

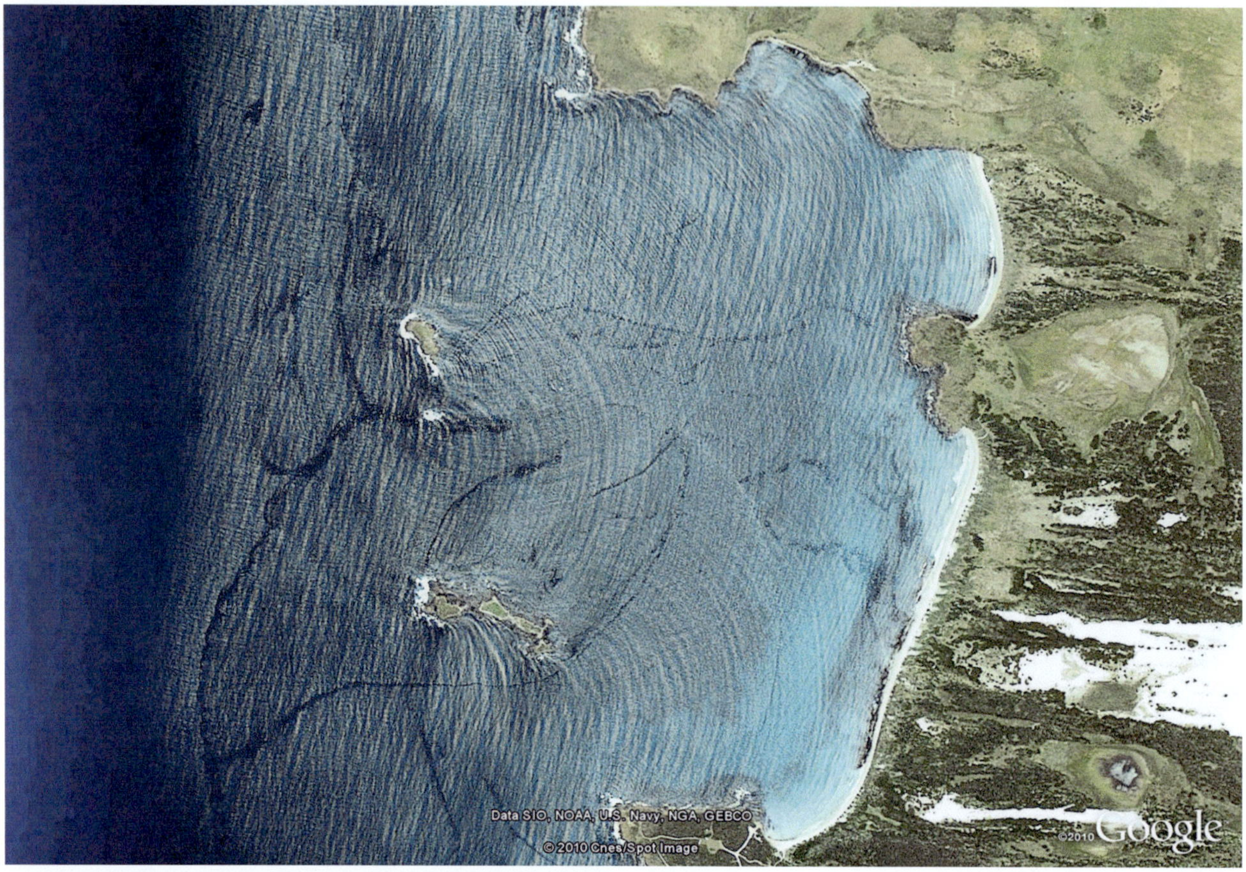
Data SIO, NOAA, U.S. Navy, NGA, GEBCO
© 2010 Cnes/Spot Image
©2010 Google

1.35a

Image © 2010 TerraMetrics
Image © 2010 GeoEye
©2010 Google

1.35b

Image © 2010 DigitalGlobe
©2010 Google

1.35c

Image © 2010 GeoEye
©2010 Google

1.36

Tidal forces

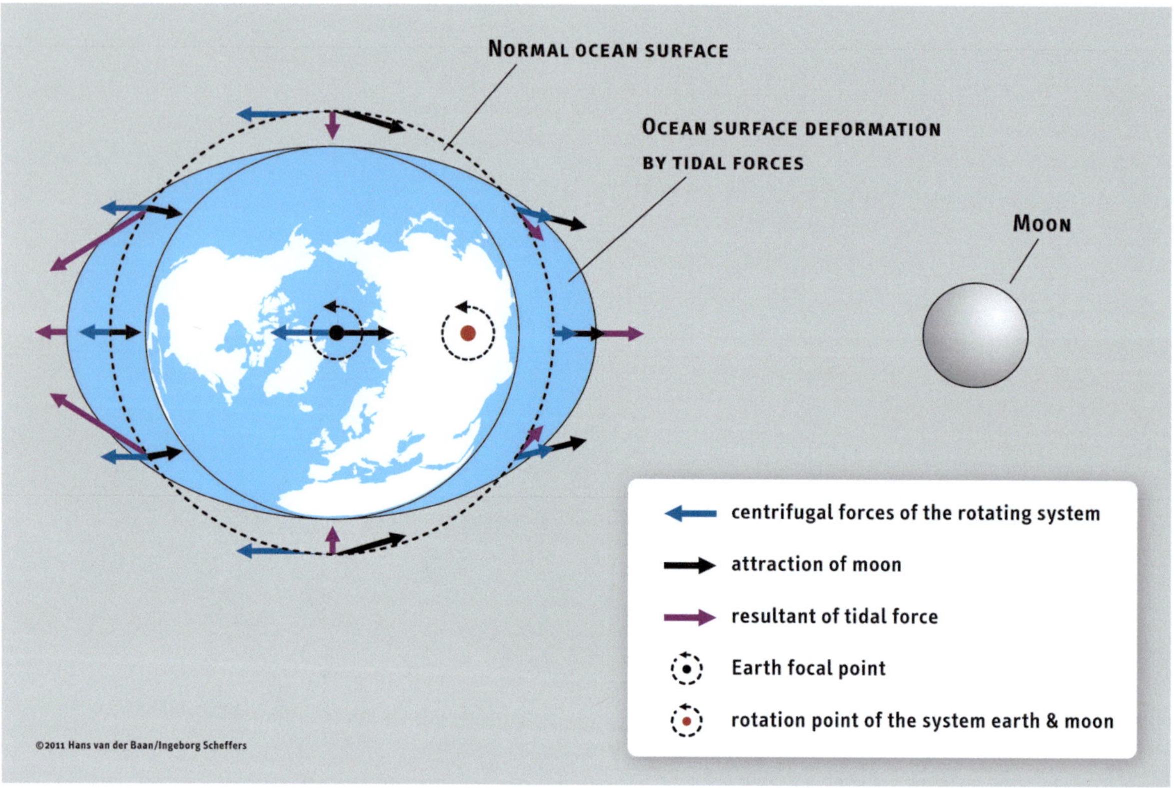

Fig. 1.37 The tides are caused by a combination of graviation forces from Earth and Moon.

Tidal curve

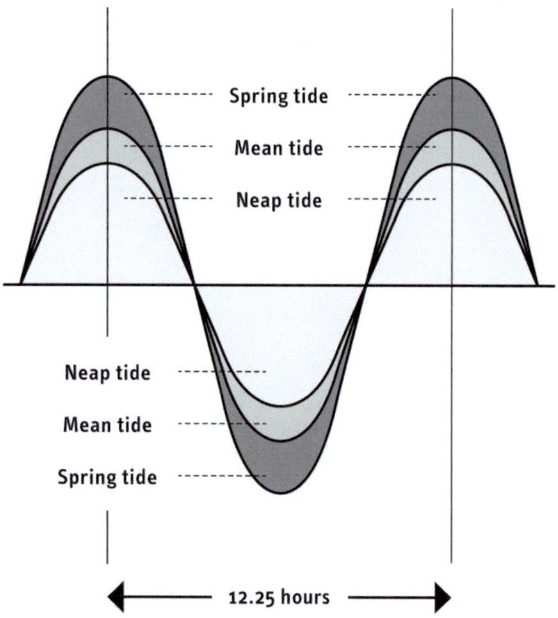

Fig. 1.38 Terms commonly used to describe a tidal curve.

1.5.3 Tides

Tides are the result of gravitational attraction and inertial forces from the Earth, the Moon and to a lesser degree the Sun and are evident as a rhythmic rise and fall of sea level from centimetres up to nearly 20 m once or twice a day (Fig. 1.37, 1.38)(Davidson-Arnott, 2010, Masselink, 2005, Schwartz, 2005; Woodroffe, 2003). Our ancestors observed this phenomenon thousands of years ago and related it to the phases of the Moon. Importantly the gravitation centre of the rotating system of Earth and Moon lies 1700 km inside the Earth due to its large mass compared to the Moon (Fig. 1.38). Both bodies turn around this point and hence we on earth see only one face of the Moon. This gravitational pull between Moon and Earth is balanced by an equal, but opposite inertia force generated by the Earth's movement (also with respect the centre of mass of the Moon-Earth-system).

Fig. 1.39 Worldwide, the highest spring tides of more than 15–20 m occur in Fundy Bay (eastern Canada). These images show a situation near high water and near low water, which in the inner part of the bay differs for 15 m or more in elevation. Length of the bay is more than 30 km. (Photo credit: NASA ASTER Science Team).

On the side of Earth that faces towards the Moon, gravitational forces as well as the centrifugal force of the rotating Earth distort the water masses of the oceans, resulting in a "tidal dome" of water, a high tide, while at the opposite side of the Earth another tidal dome is caused by centrifugal forces alone. The tidal bulges appear to move around the rotating Earth but in fact they remain stationary beneath the Moon while Earth rotates past them. When Earth, Moon and Sun (the latter with only 8% of the Moon's gravitational force) are aligned on the same axis, as during Full Moon and New Moon phases, tides of the highest amplitude are observed – the "spring tides". Tides of the lowest amplitude are experienced when the Moon and the Sun are pulling at right angles to each other and are described as "neap tides".

The tides on the open ocean are rather low, mostly less than 1 m, but if – in line with the Moon – the tidal dome moves into shallow, narrowing bays, the water may rise during a spring tide up to more than 10 m or even more than 15 m as in Fundy Bay of eastern Canada (Fig. 1.39). Tidal ranges have been classified by Davies (1980) into three groups: *microtidal* <2 m, *mesotidal* 2–4 m, *macrotidal* >4 m. Regions with high tidal ranges are the coastlines of Brittany and SW England, northern and northwestern Australia, southern Alaska, or the east coast of Patagonia (compare also Fig. 1.40). The tidal rhythm has an offset of 1–2 days after Full Moon and New Moon because of friction and inertia of the water masses. Also the phases shift for about 40 minutes from day to day, as one full circle of the Moon around Earth is not exactly 24 hours.

Theoretically every point in the ocean should show high and low water twice a day, but because of the different size of ocean basins and processes of refraction some areas show only one tidal cycle a day and others show mixed day- and half-day cycles. Combined with the tides, tidal currents occur in shallow water or in narrow gaps of the coastline. These sometimes have a velocity of many metres per second and thus have geomorphologic significance.

1.6.1 Changing Sea Levels

Sea level is defined as the mean height of water between the highest and lowest tides. Rising or falling sea level can reshape the world's coastlines and may influence some of the most densely populated areas on Earth. They include major agricultural, economic and important natural zones and cultural heritage sites. Sea-level rise is one of the most serious consequences of climate change and is predicted to rise dramatically before the end of the century. Not surprisingly, scientists want to understand sea level as thoroughly as possible. Some causes and consequences of sea level variations are well known and easy to detect, to measure and explain, but others are rather complex and need sophisticated ongoing research.

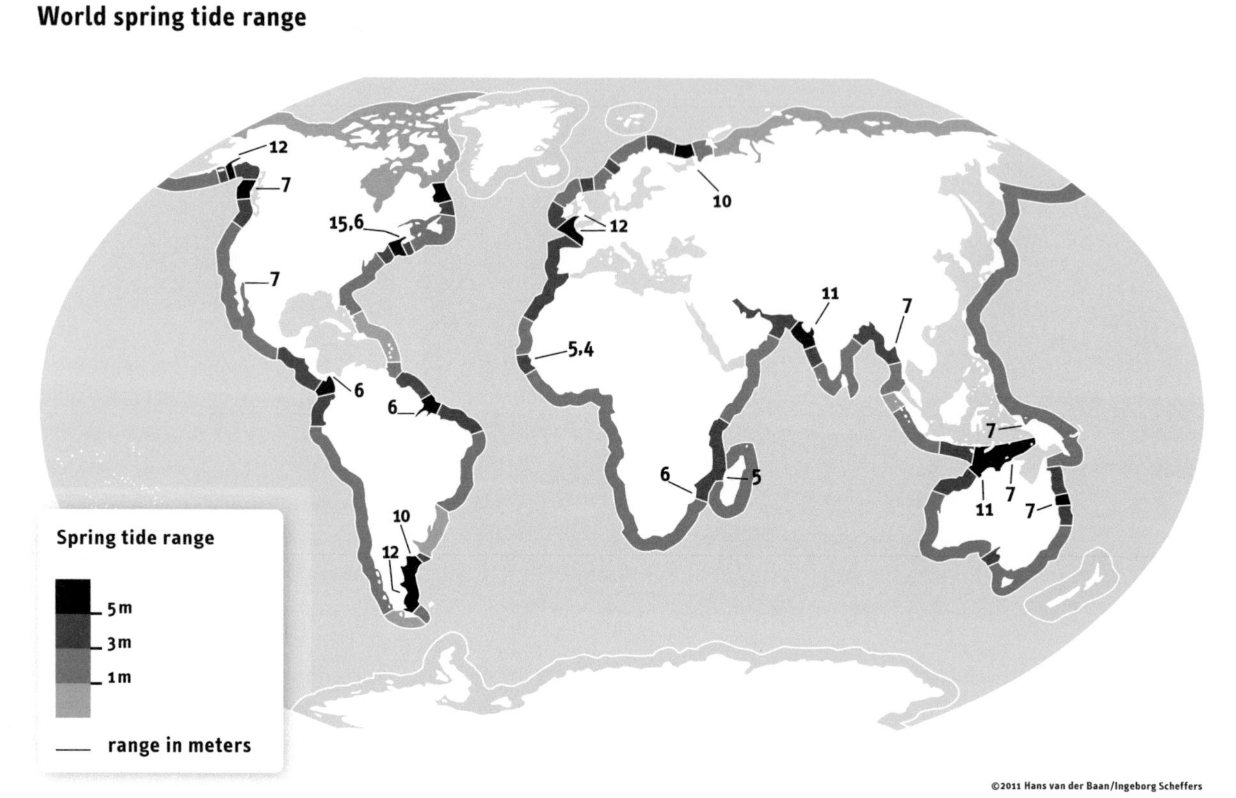

Fig. 1.40 Spring tide range around the coastlines of the world (modified from Davies, 1977).

Sea level rises and falls as the temperature and salinity of the water column varies, which is known as steric sea level change. Sea level also changes as water is added by precipitation, ice melting, or river runoff, or if it is removed by evaporation or conversion to ice (Milne et al., 2009; Pirazzoli, 1991, 1996). The greatest changes in sea level occur over long geologic time intervals when changes in the balance between ice and water on Earth cause changes in the water volume of the oceans, as for example when glaciers and continental ice sheets wax and wane. Over the last 2.6 million years (the Quaternary Period), fluctuations in the water-ice balance occurred due to the ice ages: during the colder (glacial) periods, large continental ice sheets formed at higher latitudes, withdrawing water from the oceans, and sea level decreased dramatically all over the world. During the warmer (interglacial) periods, the continental ice caps and glaciers melted and sea level rose substantially again.

Plate tectonic movements also cause the volume of ocean basins to change over Earth's history. Scientists refer to these changes that affect the ocean globally as eustatic sea level variations. Over the span of a human life, these changes are imperceptible but on geologic time spans they are key processes in the evolution of coastlines. Sea level is also affected by isostasy, a principle that envisages the lithosphere "floating" like a piece of wood or an iceberg on top of the asthenosphere. Isostasy implies that the flotational height of the object depends on its mass. The three to four kilometres thick ice sheets that formed dur-

Fig. 1.41 A beach ridge sequence deposited during a continuous glacio-isostatic uplift in northern Norway (Photo credit: D. Kelletat).

Mapping of glacio-isostatic uplifted systems of pebble beach ridges from the Varanger peninsula, Norway

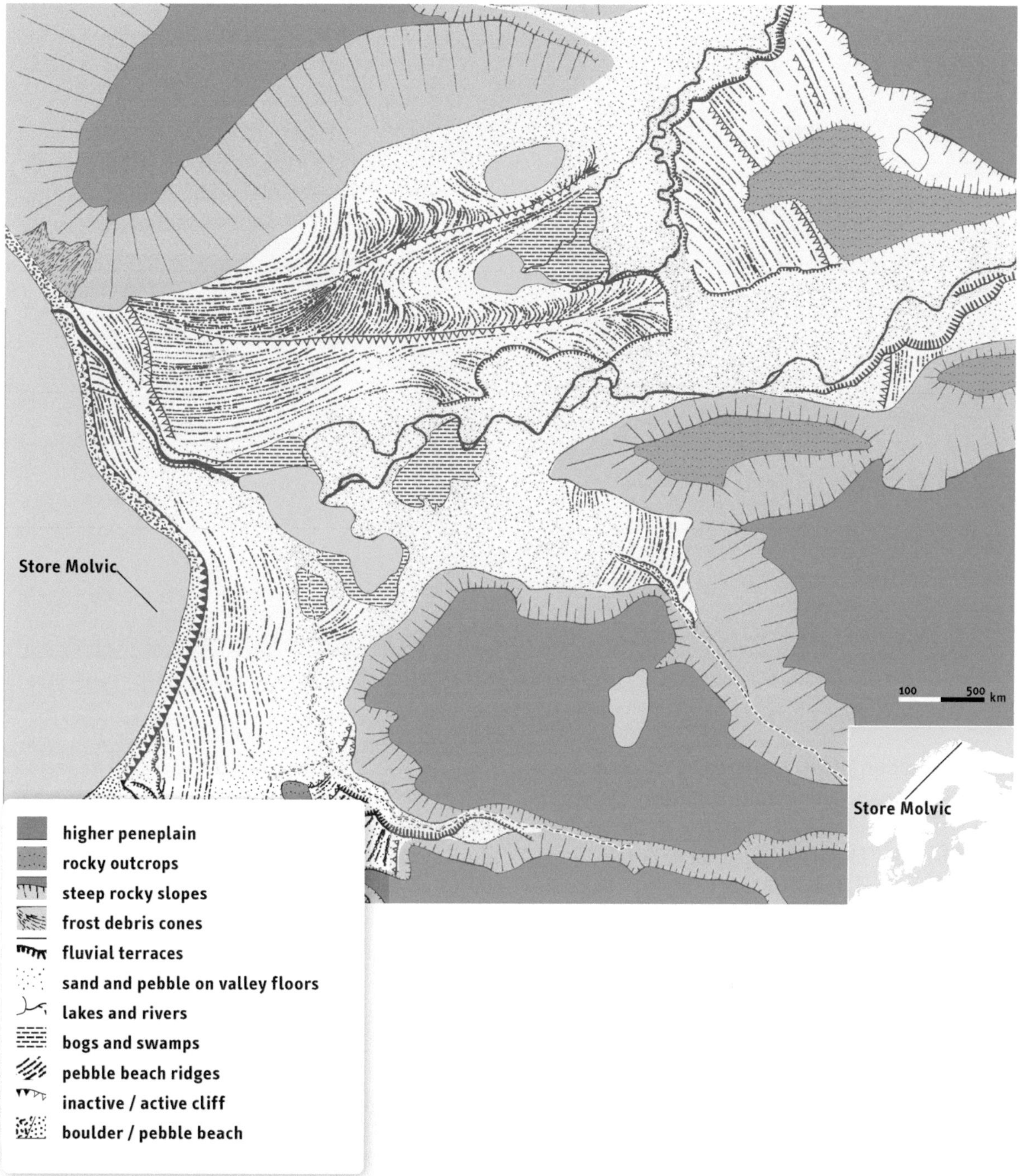

Fig. 1.42 Map of glacio-isostatic uplifted pebble beach ridges from the Varanger peninsula, northernmost Norway with ages of 11,000 to about 8,800 BP (Kelletat, 1985).

ing the ice ages provide one great example. Due to their great weight on the land masses the continental ice caps caused the Earth's crust to sink for several hundred metres resulting in the underlying asthenosphere flowing outward. As a consequence sea level rose relative to the land. When the ice melted, the land mass rose once more and the asthenosphere began to flow back into the underlying areas, resulting in recent uplift of these formerly glaciated regions.

Fig. 1.43 At Hudson Bay in Canada numerous glacio-isostatic uplifted beach ridges have been formed over time. Width of scene is about 30 km at *58° N and 93° W.* (Image credit: ©Google Earth 2010)

This phenomenon of isostasy may last over many thousands of years after melting of the ice sheets and can still be observed occurring in some areas of the Northern Hemisphere today. For example shorelines around the Northern Baltic Sea, the Scandinavian land mass or the shorelines around Hudson Bay, North America are still rising up to 1 cm/y in the regions of the thickest former ice load. Shorelines deposited about 10000 years ago, when global sea level was about 50 m lower, can now be found up to 200 m above sea level (Figs. 1.41–1.43, see also Kelletat, 1985; Pirazzoli, 1996).

Fig. 1.44 Uplifted dead coral in Simeulue, Indonesia, from the Boxing Day earthquake on 26 December 2004. (Photo credit: Craig Shuman, Reef Check/Marine Photobank).

Fig. 1.45 Notch and double algal rim at +7 m asl in western Crete, Greece, uplifted on July 21st, 365 AD (Photo credit: D. Kelletat).

Fig. 1.46 The uplift of western Crete in 365 AD has exposed these cup-shaped micro-atolls of vermetids and calcareous algae (Photo credit: D. Kelletat).

Fig. 1.47 A similar phenomenon of several bio-erosive notches in limestone on the east coast of Rhodes island (Greece) (Photo credit: D. Kelletat).

While the effect of glacio-isostasy is confined to formerly glaciated areas and their vicinity, isostatic movements of the Earth's crust are also caused by changes of mass due to the wearing down of mountain ranges and the deposition of sediment loads on the ocean floor as weathering and erosion transport particles from land into the sea over tens of millions of years. On a smaller scale, regional subsidence due to sediment load on the sea floor takes place in nearly all large deltas of the world. Here, large thicknesses of sediment accumulate in a rather short time (some thousands of years) on a small part of the ocean crust resulting in coastal submergence. The associated relative sea level rise may reach many meters within a thousand years.

Tectonic movements of the land may cause uplift or subsidence and thus local emergence or submergence of the coastline. For example, earthquakes may be accompanied by a vertical motion as either uplift or subsidence. Two of the largest earthquakes ever measured in the instrumental record have been the Sumatra-Andaman earthquake of 26/12/2004 which had a magnitude of 9.1 on the Richter scale and the Japan Tohoku earthquake on 11/3/2011 with a magnitude of 9.0. (The largest measured earthquake occurred in the subduction zone off the coast of Chile in 1960 and had a magnitude of 9.5 while the Great Alaska earthquake in 1964 had a magnitude of 9.2). During the Andaman-Sumatra earthquake islands west of Sumatra were uplifted by 1–1.5 m, and former flourishing fringing coral reefs are now exposed above sea level (Fig. 1.44). The devastating earthquake off Japan's coast in March 2011 caused subsidence along the East coast of Honshu (Japan's main island) with a maximum subsidence of 75 cm, while the horizontal displacement has been of up to ca. 4.4 m eastwards.

Another well-studied historical example is uplift of the western parts of Crete and Rhodes Island (Greece) during a massive earthquake on July 21, 365 AD. Within seconds the coastline of Western Crete was uplifted for as much as 9 m (Figs. 1.45 to 1.47). Evidence for this cataclysmic event are visible today in the coastal landscape in the form of notches or micro-atolls which formed at sea level and are now well above the present day shoreline.

Beside these sudden events, vertical tectonic movements along the margins of converging plates over longer geologic time spans may have uplifted former beaches or coral reefs to positions far above modern sea level. This is the case for the Huon Peninsula in Papua New Guinea, Barbados, Haiti or parts of Cuba where flights of uplifted coral reef terraces occur many hundreds of metres above sea level. Because tectonic movements and eustatic sea level fluctuations may occur in the same or opposite directions, at different rates or simultaneously, it can be a very challenging exercise for scientists to understand the sea level history of a certain coast.

Sea-level curve of the Pleistocene

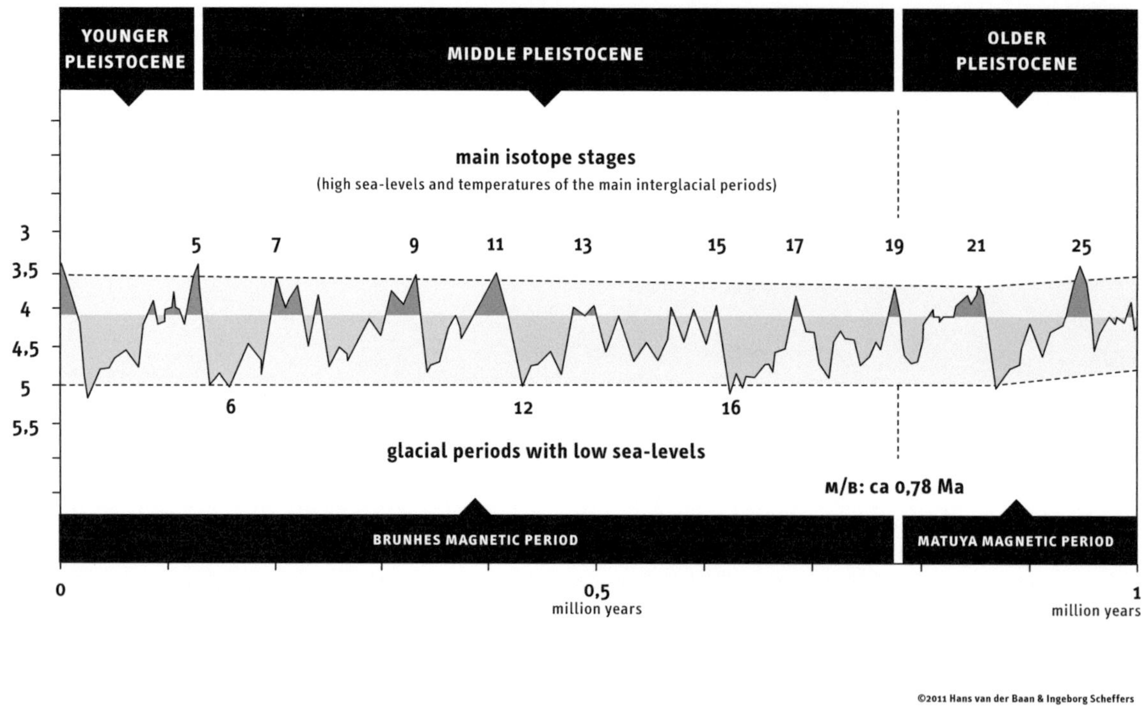

Fig. 1.48 The typical saw-tooth curve of glacio-eustatic sea-level movement between ice ages and warm phases during the last 2 million years (acc. to Shackleton, 1995).

Fig. 1.49 Evidence of a former lower sea level is present in this fresh water karstic spring (smooth place in the water) in the Bay of Argolis, Peloponnese, Greece. Karst solution forms were developed down to sea levels of glacial times, which were more than 100m lower than today. (Photo credit: D. Kelletat).

Fig. 1.50 Caves in eastern Sardinia (Italy) show two smooth carved notches as evidence of formerly higher relative sea levels. (Photo credit: D. Kelletat).

Fig. 1.51 A notch has been carved by pebbles moving in the surf about 125,000 years ago on the south coast of Crete, Greece. (Photo credit: A. Scheffers).

Fig. 1.52 A sharply incised bio-erosive notch in coral limestone from a former sea level highstand (warm phase 125,000 years ago), Bonaire, southern Caribbean. (Photo credit: S. Scheffers).

Fig. 1.53 The horizontal cover on the sloping strata on Ibiza Island are beach deposits from a former interglacial (warm) sea level phase. (Photo credit: D. Kelletat).

Fig. 1.54 Rounded pebbles with shells (gastropods) document an old beach deposit under younger angular rock fall debris (Nauplion, Greece). (Photo credit: A. Scheffers).

1.6.2 Sea level changes during the Ice ages

Most dramatic sea level changes took place during the Quaternary Period, or the last 3 million years in Earth's history when the climate fluctuated between ice ages and warm phases (Fig. 1.48). The last warmer (interglacial) period comparable with the Holocene, that is the current warm climatic period, occurred between 130,000 and 118,000 years ago. At that time, sea level was 4 to 6 metres higher than it is today. This interglacial phase was followed by

Fig. 1.55 Shell layers along creeks in the Pampa near La Plata, Argentina. (Photo credit: D. Kelletat).

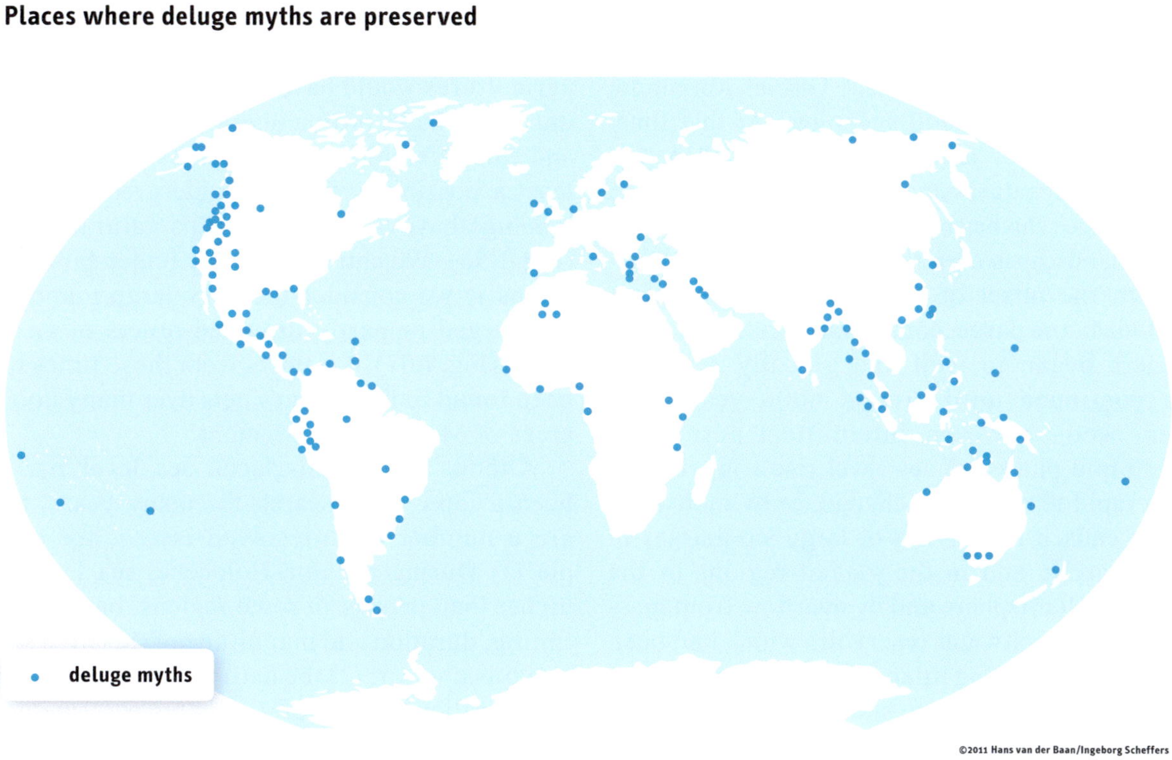

Fig. 1.56 Places on earth where deluge myths are preserved.

Fig. 1.57 The Serapeum, a temple from late Roman times near Pozzuoli in Italy, now located at groundwater level. The columns of the left part of the picture show dark rings with marine borings from a time when the sea inundated the temple. As it certainly was built on dry land, it provides evidence of several yo-yo-like earth movements during the last 2,000 years in this region. (Photo credit: D. Kelletat).

an irregular transition into the last colder (glacial) period, with the growth of ice sheets and fall of sea level culminating in the Last Glacial Maximum (LGM) 21,000 to 18,000 years ago. At that time, sea level was 121 to 125 metres lower than it is today. The signatures and fingerprints of these sea level changes can be observed and deciphered in our coastal landscapes even today (Figs. 1.47–1.55).

With the onset of a warming climate after the LGM, the large continental ice sheets and glaciers began to melt and globally sea level rise continued until around 6000 years ago. There were, however, intermittent periods of more rapid phases of sea level rise triggered by more rapid temperature change, or by meltwater pulses caused by calving of large ice masses in the Antarctic and in the glacial regions of the Northern Hemisphere and by overflow from massive natural meltwater reservoirs which had been formed by retreating inland glaciers. Overall, the postglacial sea level rise started slowly, accelerated in the mid part of this time span and slowed down again by about 8000 BP. Consider that in a time span of about 8000 years sea level rose by 120 m, which equates to 15 mm/year on average or a rate of approximately 1°/1.5 m per century, with occasional sharp increases. During the two periods of melt water pulses, each lasting around 300 years, sea level rose by 5 metres per century! Beside the average rates of rise, there have been periods that lasted for centuries during which sea level rose ten to twenty times faster corresponding to a sea level rise of many centimetres/year. Nearly all coasts have thus experienced submergence or "drowning" and evidence for lower glacial sea levels is found seaward of present coastlines out to depths of 100 m or more. A terrific visualization is the Google Earth Blue Marble 3000, developed at the Zurich University of Applied Sciences by Adrian Meyer and Karl Rege. It shows the earth starting at the last glacial maximum 21000 years ago and ends 1000 years in the future (for reference visit: Blue Marble 3000; http://radar.zhaw.ch/bluemarble3000_en.html).

In flat and low lying coastal areas with developed cultures such as in Mesopotamia and the deltas of the Nile, Indus and the Yellow River, the postglacial sea-level rise resulted in a progressive inundation of the land which may have encroached many hundreds of metres or even kilometres per year. During a number of generations areas which were formerly use for coastal living, hunting and the development of primitive agricultures would have thus been drowned. This must have had a profound effect on the Mesolithic and early Neolithic cultures and societies. It is at least a possibility that the widespread myths of a deluge have their roots in this natural process (Fig. 1.56). We can more easily understand these myths if we consider the very large number of submerged or partly drowned relicts of ancient times (Fig. 1.57). Artefacts from those times have been found in fisherman's nets over many flooded areas of Mesolithic settlement.

Although this post-glacial sea level rise has been a topic for research for many years, there are a number of unresolved issues, for example: (1) During the mid-Holocene, sea level was higher than present in some regions, but the exact timing, duration and magnitude of this high stand remains unclear; (2) the nature of sea level changes on millennial scales – debates exist between models of a smoothly falling sea level since the mid-Holocene high stand or a step-like recession; and (3) sea-level variations on shorter time scales, e.g. at decadal to century resolution relevant to human society and usable for risk assessment purpose.

Let us briefly consider that sea level may also change within seconds or hours during short-period sea level variations caused by waves, tides or storm surges. Irregular but frequently occurring variations in sea level are triggered by air pressure: under an atmospheric high pressure system sea level will be lowered, whereas sea level is higher than normal in the centre of a low pressure system like a cyclone. Variations of one hectopascal (hPa) of pressure normally results in 1 cm of sea level (rise or fall). This is true for the open ocean as well as for the coastline, but at the coastline the effect may be stronger and more dangerous. During the passage of a hurricane over the shoreline not only low pressure raises sea level, but also wind forces push water towards the coast, producing a storm surge. During Hurricane Katrina in 2005 the height of the associated storm surge reached around +7 m in the Gulf of Mexico along the US south coast.

Tsunami waves, only decimetres high in the open ocean and triggered by earth movements, volcanism or massive slumping, may reach considerable run-up levels at the coast of 20 m or even more. The highest run-up during the 2004 Indian Ocean Tsunami was 51 m above sea level at one point in north-western Sumatra,but around 30 m along several hundred kilometres of coastline. We will look more closely at marine natural hazards in Chapter 8 and discuss more recent and future sea level rise in Chapter 9.

References

Blue Marble 3000. http://radar.zhaw.ch/bluemarble3000en.html).

Couper A (Ed., 1983) The Times Atlas of the Oceans. Times Books Limited, London.

Davidson-Arnott R (2010) Introduction to Coastal Processes and Geomorphology. Cambridge University Press, Cambridge.

Davies JL (1980) Geographical Variation in Coastal Development. 2nd, edn. Longman, Edinburgh-London.

Kelletat D (1985) Studien zur spät- und postglazialen Küstenentwicklung der Varanger-Halbinsel, Nord-Norwegen. Essener Geographische Arbeiten 10:1–110.

Kelletat D (1999) Physische Geographie der Meere und Küsten. Teubner Studienbücher Geographie, 2 ed..

Masselink G (2005) Waves. In: Schwartz M (ed) Encyclopedia of Coastal Science 1069–1074. Springer, Dordrecht.

Milne GA, Gehrels WR, Hughes CW, Tamisiea ME (2009) Identifying the causes of sea-level change. Nature Geoscience 2:471–478.

Paskoff R, Oueslati A (eds. 1991) Modifications and coastal conditions in the Gulf of Gabes (Tunisia) since classical antiquity. Annals of Geomorphology, NF Suppl 81:149–162.

Pirazzoli PA (1991) World Atlas of Holocene Sea-Level Changes. The Netherlands: Elsevier Oceanography Series 58, Amsterdam.

Pirazzoli PA (1996) Sea-level changes: the last 20.000 years. Wiley, Chichester.

Schwartz ML (Ed., 2005) Encyclopedia of Coastal Science. Springer, Dordrecht

Seibold E, Berger WH (1982) The Sea Floor. An Introduction to Marine Geology. Springer, Berlin-New York-Heidelberg.

Shackleton NJ (1995) New Data on the Evolution of Pliocene Climatic Variability. In: Vrba ES, Denton DH, Partridge TC, Burckle LH (ed) Palaeoclimate and Evolution, with Emphasis on Human Origin. Yale University Press, London.

Vött A, Brückner H, Brockmüller S, Handl M, May SM, Gaki-Panpanastassiou K, Herd R, Lang F, Maroukian H, Nelle O, Papanastassiou D (2009) Traces of Holocene tsunamis across the Sound of Lefkada, NW Greece. Global and Planetary Change 66:112–128.

Woodroffe CD (2003) Coasts: Form, Process and Evolution. Cambridge University Press, Cambridge.

World Ocean Review (2010), retrieved 16 March 2011, http://worldoceanreview.com/en.

2

Coastal Landforms and Landscapes

ABSTRACT Coastal geomorphology is the study of coastal landforms and their evolution over time. In this topic, we set the scene for discussing the variety of coastal landforms and landscapes and the major climatic and oceanic forces that shaped them during their geologic history. First, we briefly explore landforms along ice coasts, which have an extent of some tens of thousands of kilometres, either as floating shelf ice of Antarctica or calving glacier fronts of Greenland, Alaska or Chile. We then continue with coastal forms dominated by endogenic signatures like faulting, folding or jointing, or which are of volcanic origin and partly drowned by sea level rise in the warm period of today's climate.

2.1 Classification of Coastal Landforms

Due to the complexity of coasts, numerous attempts to organize coastal features or processes have been put forward but hitherto no single system of classification has been comprehensive in scope or coverage (Finkl, 2004; Kelletat, 1999; Woodroffe, 2003). An example of a hierarchical classification system of natural coastlines with respect to their genesis and evolution of coastal features is given in table 2 (Read the table from left to right as it defines coastlines in very broad terms on the left hand site and in more detail at the right). In the widest sense, every coastline on Earth is either advancing or retreating (it may also be stable, but only for a very short period of time), or has advanced or retreated. The difference is evident: the terms *advanced* or *retreated* describe the history of evolution of the coastal forms; the terms *advancing* or *retreating*, the present day status. The next category of classification is defining *why* a coast is advancing or retreating. A coast may advance towards the sea because of emergence. For example tectonic forces within the Earth's crust may uplift a coastline over geologic time. Alternatively constructive and/or depositional processes such as the growth of coral reefs or the deposition of sediments in delta regions may result in regression, or thirdly, it may result from sea level fall. A retreating coast is the result of relative sea level rise where tectonic movements cause a sinking of the crust or rise of sea level drowns the coastline. So too it may be affected by abrasion, which is the general term for coasts experiencing natural destruction and erosion by different forces. A more detailed differentiation classifies coastlines according to the processes which shaped them in the recent past and continue today: For example, constructive coastlines built up by organic processes (think again of coral reefs or visualize a coastline where extended mangrove flats or sea grass pastures capture sediments and extend the shoreline laterally over time), or new land along a shore built up by lava flows. The classification of coastlines

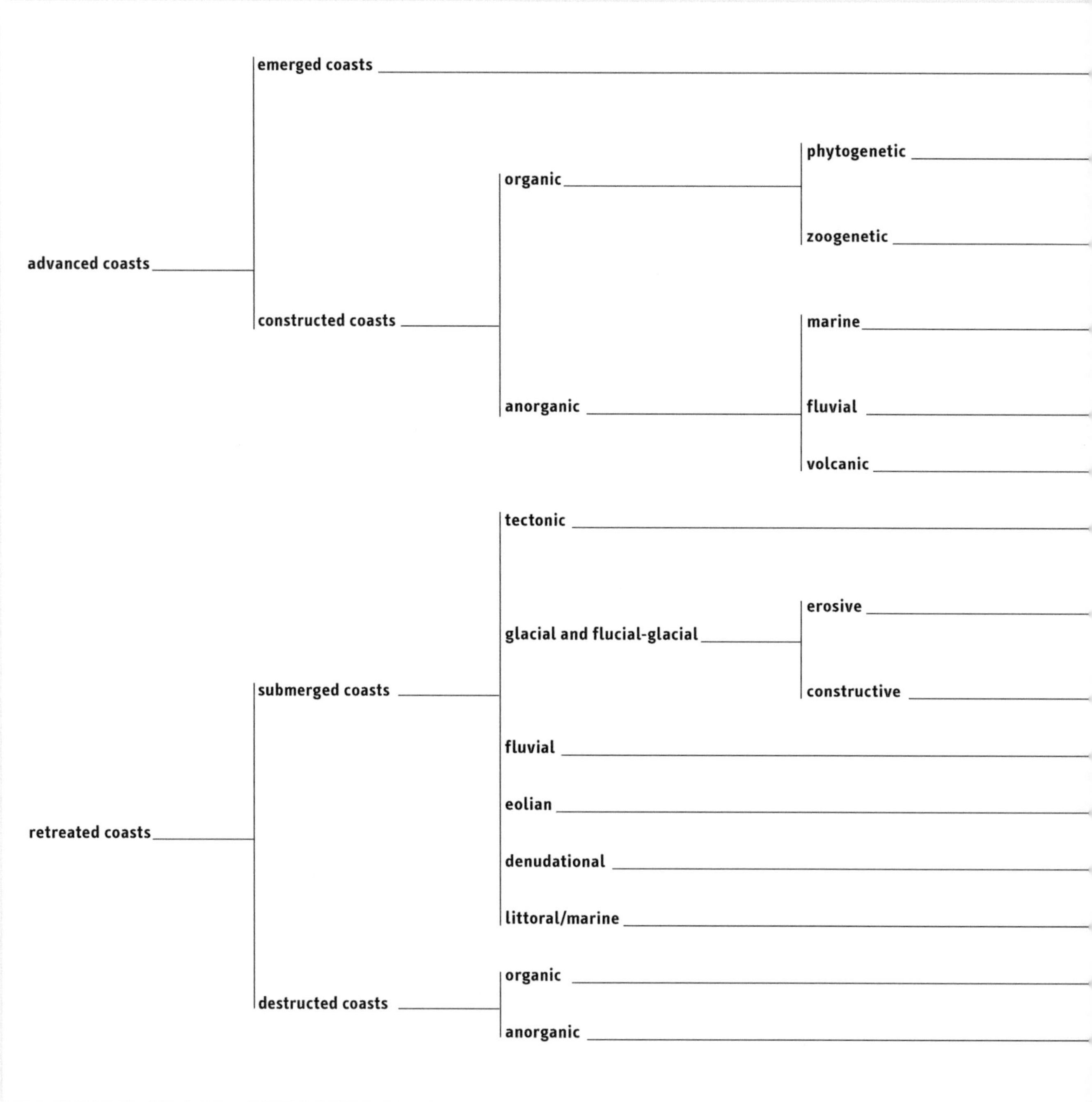

Table 2 Genetic classification of coastlines (based on Valentin, modified by D. Kelletat 1995).

by Kelletat (1995) recognizes numerous categories of advanced and retreated coastlines based on a range of criteria, (see below). In the following topics, we present different landscapes inundated by the sea and discuss the resulting coastal landforms. We start with coastlines in a dynamic environment of magnificent natural scenery at the edge between the great continental ice caps and the sea – ice coasts.

2.2 Ice Cliffs, Calving Glaciers and Sea Ice

Only about 10 percent of Earth's land surface is now covered by glaciers, almost all of this is blanketed by the huge ice sheets of Greenland

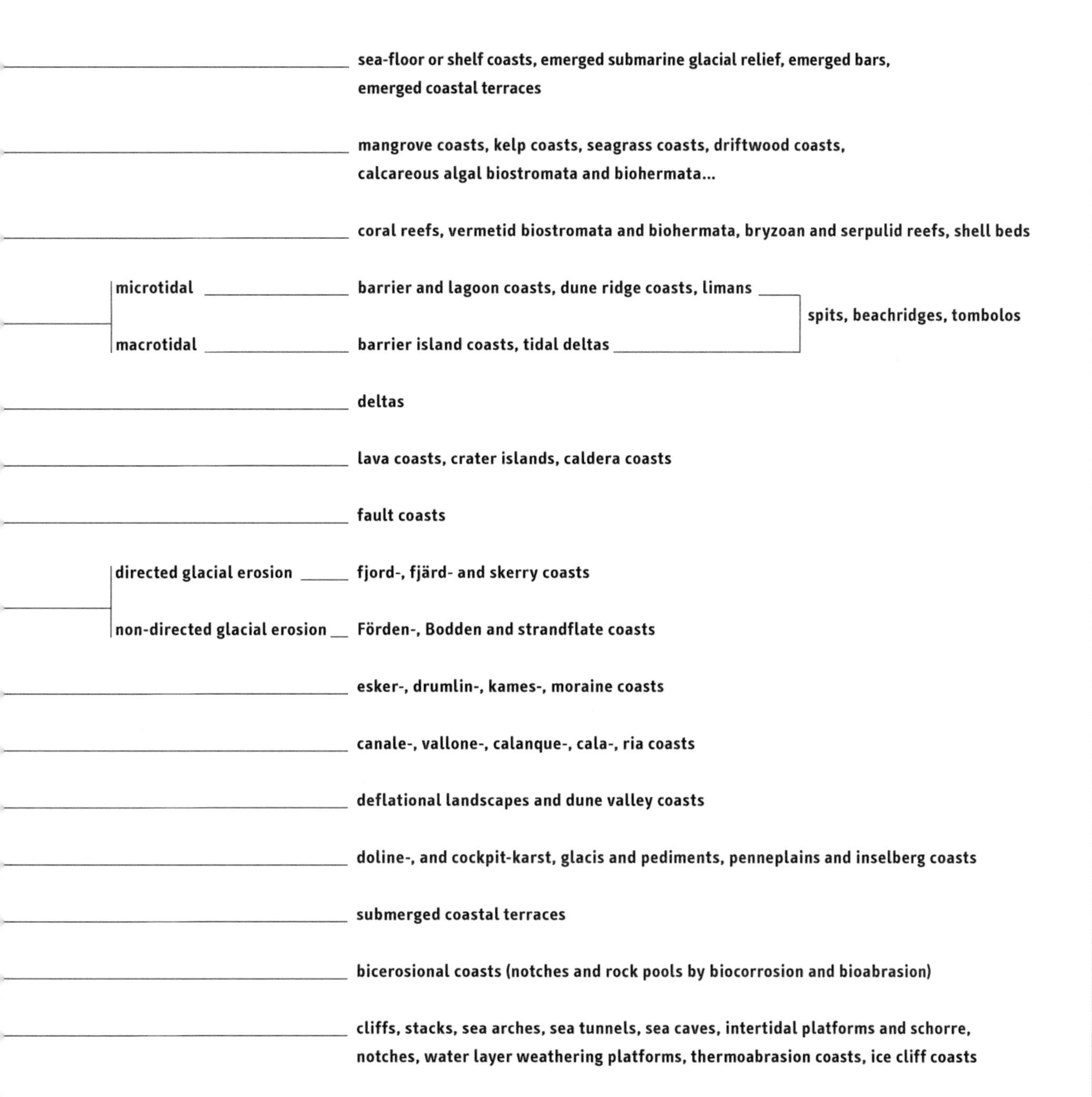

and Antarctica. However, as recently as 20000 years ago, ice sheets covered almost three times more land than they do now. The landscapes and coastlines of many continents have been sculpted by glaciers now melted away. To a geoscientist, a block of ice is a rock, a mass of crystalline grains of the mineral ice which has some unusual properties. For example ice is less dense than water, which is why icebergs float in the ocean. Because ice is deformable, it flows readily downhill like a viscous fluid. Glaciers are large masses of ice on land that show evidence of being in motion or of once having moved under the force of gravity. In mountainous regions, glaciers have eroded steep-walled valleys, scraped bedrock surfaces, and plucked huge blocks from their rocky floors. During the ice ages, glaciers pushed across entire continents, carving far more topography than

Fig. 2.1 Shelf ice in Antarctica forms impressive coastlines over thousands of kilometres. The ice cliffs above water are about 30 m high while the entire thickness of the front edge is nearly 300 m. (Photo credit: Josh Landis, National Science Foundation).

rivers and wind. Glacial erosion creates enormous amounts of debris, and ice transports huge tonnages of sediments, depositing them at the edges of glaciers, where they may be carried away by meltwater streams. Glacial processes affect the water discharge and sediment loads of major river systems, the erosion and sedimentation of coastal areas, and the quantity of sediment delivered to the oceans.

Ice and glaciers form spectacular coastlines stretching over thousands of kilometres in colder, high-latitude climates. Today, the world's largest ice sheets overlay much of Greenland and Antarctica. The glacial ice of Greenland and Antarctica is not confined to mountain valleys but covers virtually the entire land surface. The upper surface of an ice sheet resembles an extremely wide convex lens. From this central area, the ice surface slopes to the sea on all sides. Though very large with a thickness of around 3000m at its highest point, the Greenland icecap is dwarfed by the Antarctic ice sheet. Here, ice blankets 90% of the Antarctic continent, covering an area of about 12.5 million square kilometers and reaching thicknesses of 4000 m. Overall, 11000 kilometres of Antarctica's coastline are rimmed by ice shelves floating on the ocean with ice cliffs fronting the main glaciers on land. The best known of these is the Ross Ice Shelf that floats on the Ross Sea. The ice cliffs of Antarctica (Fig. 2.1–2.2) are dynamic: shelf ice may break off in huge segments up to several thousands of square kilometres – a process called iceberg calving – and float into warmer regions as tabular ice bergs. In the Northern Hemisphere valley glaciers may flow down coastal mountain ranges and terminate at the ocean's edge where they calve with irregular iceberg forms of much smaller size. This is typical

▸Fig. 2.2 Disintegration of an Antarctic shelf glacier into tabular ice bergs. Image shows a 7 km wide section at *70° 30′S* and *8° 22′W.* (Photo credit: ©Google Earth 2010).

▸Fig. 2.3 Antarctic shelf ice coastline. Tabular ice bergs are caught in thick sea ice, which breaks off in spring time. Image shows 17 km of the Antarctic coastline at about *66° 41′S and 89° 55′E.* (Photo credit: ©Google Earth 2010).

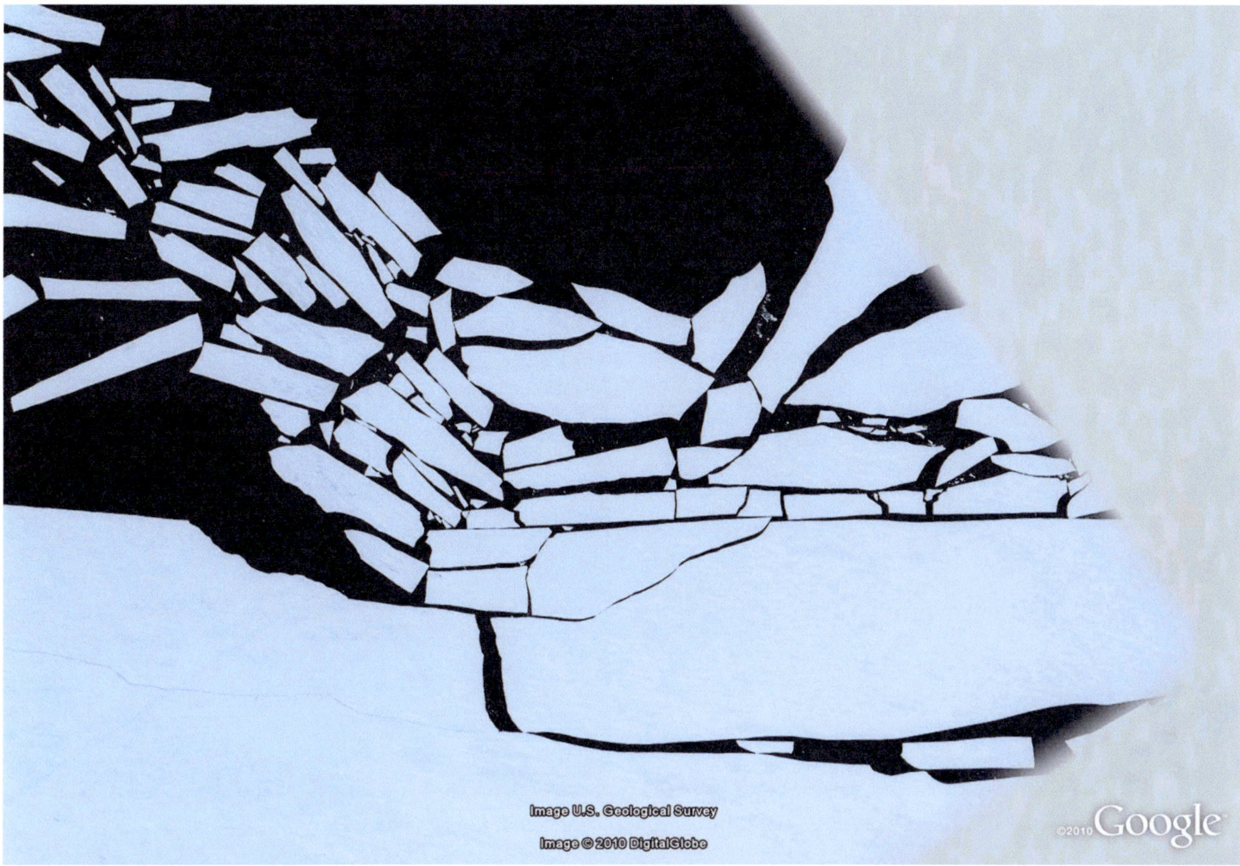
Image U.S. Geological Survey
Image © 2010 DigitalGlobe
©2010 Google

2.2

Image © 2010 DigitalGlobe
©2010 Google

2.3

Fig. 2.4 Typical "calving" of an Arctic outlet glacier in Greenland (Photo credit: iStockphoto LP).

of Greenland, Spitsbergen, Novaja Semlja, southern Alaska, or southern Chile (Figs. 2.4–2.7).

Most of the Antarctic's glacial shrinkage or loss of ice mass results from warming and melting taking place at the glacier's leading edge and at its base. Thus, even though a glacier is moving downward or outward from the centre of a continental icecap, the ice edge may be retreating. Glacier retreat by calving may reduce the floating ice in a very short time and can be incredibly spectacular to witness, with massive ice slabs splashing down into the ocean creating enormous waves if breaking off mountain glaciers, whereas the shelf ice just separates and floats away. These newly formed icebergs gently float away, driven by currents or the wind.

Using high-resolution radar satellite mapping, glaciologists have observed that several Antarctic glaciers have retreated more than 30 km in just 3 years. Over the past 20 years or so, enormous pieces of ice have snapped off Antarctic glaciers. In March 2000, an iceberg of slightly less than 10.000 km^2 calved from the Ross Ice Shelf. More recently, in February and March of 2002, a portion of the Larsen Ice Shelf of about 3250 km^2 shattered and separated from the east side of the Antarctic Peninsula. The fracturing of this piece of the ice sheet produced thousands of icebergs. In Glacier Bay, Alaska, a system of valley glaciers has retreated for about 120 km during the last 200 years, and in Jacobshavn Fjord of western Greenland (Fig. 2.6 and 2.7), where the largest number of ice bergs in the Northern Hemisphere are produced, a 5–6 km wide ice flow has retreated by calving at a rate of 0.5–1 km/year, but the icebergs still fill a large portion of the fjord.

Icebergs may become unstable as they undergo several cycles of melting and freezing on their journey into warmer latitudes or are affected by warm water currents. They may also tilt or turn over as indicated by inclined melt notches in exposed parts of the icebergs tilted from their former horizontal position. Or they even can explode! Dr. Gregory Stone, a member of a National Geographic expedition, described the incident in his book, "Ice Island": "The enormous iceberg ...heaved upwards, one end pausing high in the air like the bow of a foundering ship, then crashed down, creating waves that swept through all of Hallett Bay and rocked our boat...[it] rose one last time and seemed to explode into millions of pieces like shards of crystal, covering two square miles of ocean. Later, we circled

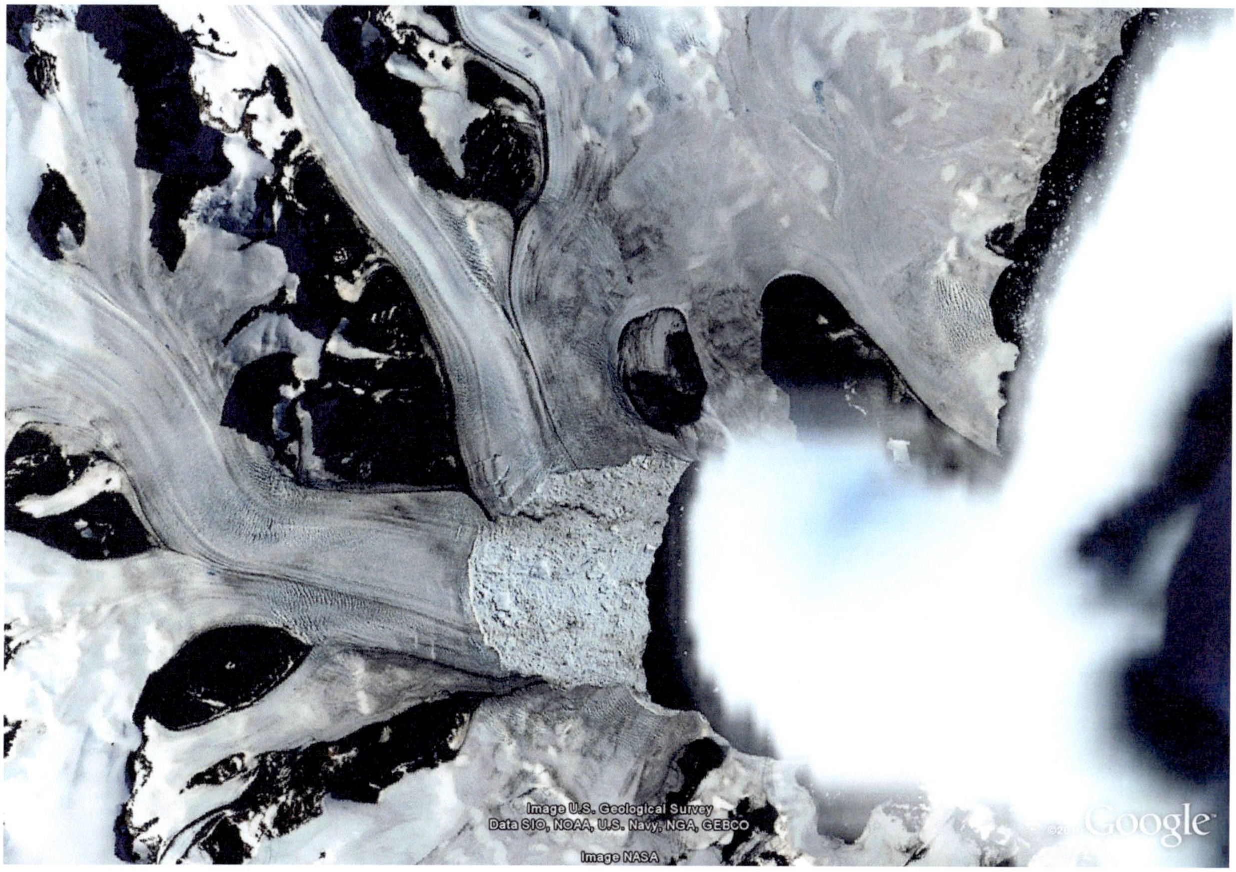

Fig. 2.5 Arctic outlet glaciers from eastern Greenland producing icebergs of irregular forms at *64° 11′N* and *59° 06′W*. Width of scene is 45 km. (Photo credit: ©Google Earth 2010).

Fig. 2.6 Glacier calving in west Greenland. The image shows a small section of the up to 10 km wide and more than 40 m high calving front of Jacobshavn Icebre near Illulisat, Disco Bay, western Greenland where most of the northern hemisphere icebergs (many thousands per year) are produced. Due to a drowned terminal moraine at the outlet of the fjord, now 60 km apart from the calving front, large icebergs touch ground here and stop the outflow into Disco Bay for a while.(Photo credit: D. Kelletat).

Fig. 2.7 Drifting ice bergs in Jacobshavn Fjord near Illulisat, western Greenland, all calved from the main outlet glacier. The height above water reaches up to 40 m in this area (Photo credit: D. Kelletat).

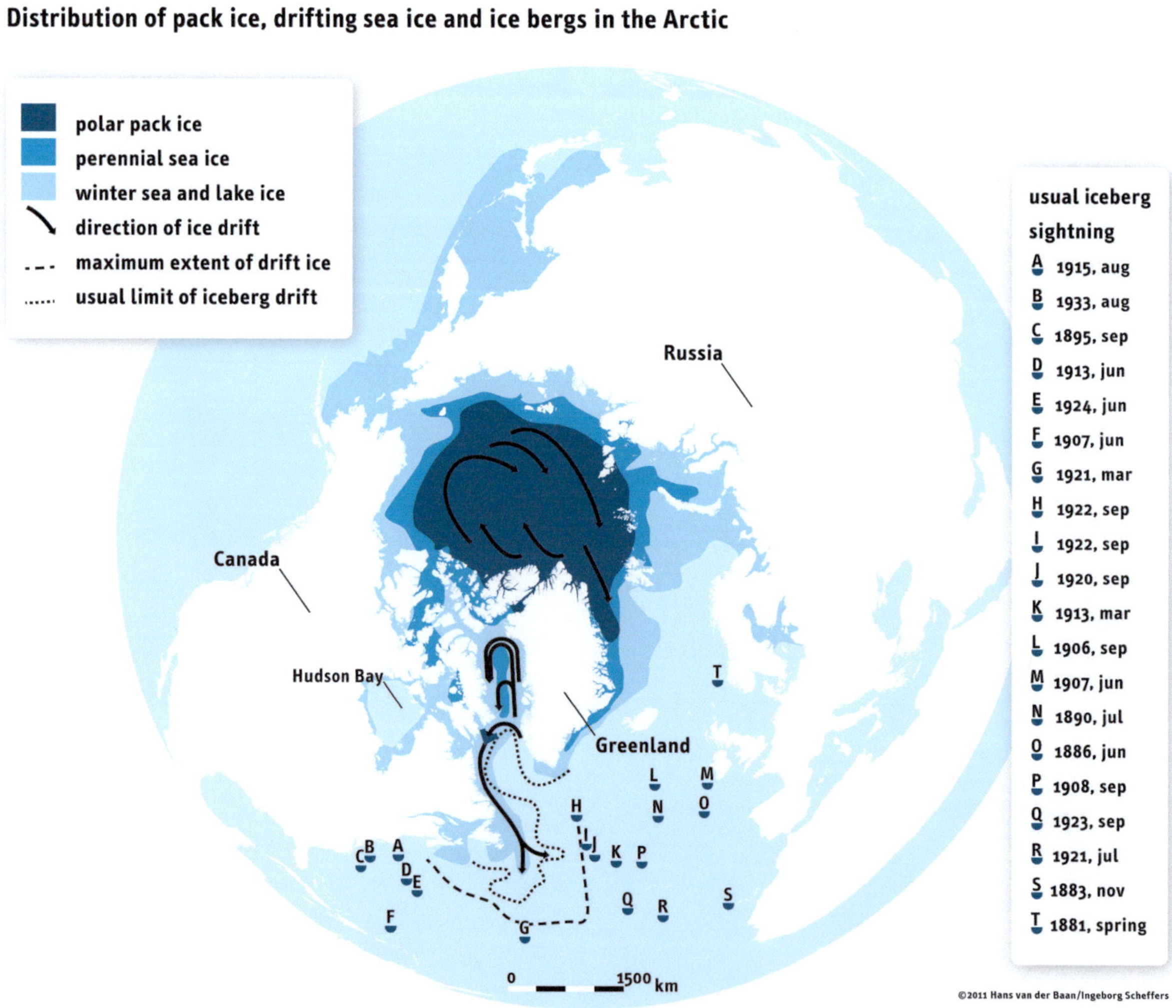

Fig. 2.8 Distribution of pack ice, drifting sea ice and ice bergs in the Arctic.

Fig. 2.9 Sea ice forms during the winter months. By collision of broken pieces of this floating ice a kind of "pancake ice" develops. Drifting sea ice may transform depositional coastlines and push debris landward or sideways. (Photo credit: Zee Evans, National Science Foundation).

the debris field of shattered ice." Icebergs can also become erosional agents ploughing deep gouges in the nearshore sea bottom, because once in motion due to wind or currents they are hard to stop!

Sea ice floats as a thin veneer on the polar oceans; it is vast in its extent, but comprises only about 1/1000 of Earth's total ice volume (Fig. 2.8 for the Northern Hemisphere). Sea ice is in constant motion driven by wind and ocean currents. It is also actively forming coastal environments and landscapes. Ice formation and ice movement is quite different in the Arctic and Antarctic Oceans: The Arctic Ocean is surrounded by land and sea ice lasts over several years or decades without melting whereas most Antarctic sea ice (~ 85%) is annual or first year ice. During the Northern Hemisphere summer virtually all Arctic coasts are free of sea ice for varying lengths and time. Exceptions are northern Greenland, parts of the Canadian Arctic archipelago and Ellesmere where sea ice may last throughout the year.

Sea ice as a geologic force can act both as an agent of erosion and of protection. One impact of sea ice is that, if close enough to the shore, it dampens waves and reduces their effect on the coastline. An important process is the movement and dislodgement of sediments within the sea ice, either offshore, onshore or along the coast with coast-parallel longshore currents. The movement of sea ice on the beach or near the shore can produce scour marks or ridges in foreshore areas especially if larger boulders of rock are incorporated in the ice matrix. If large rocks, e.g. boulders from outwash moraines, are pushed ashore, they may remain and form so-called ice pushed ridges or boulder barricades. Sediments incorporated into the ice may polish intertidal rock formations, forming for example the famous boulder pavements of the St. Lawrence estuary in eastern Canada.

Ice forming on the sea may show many different structures and variable ages. There may be

2.10

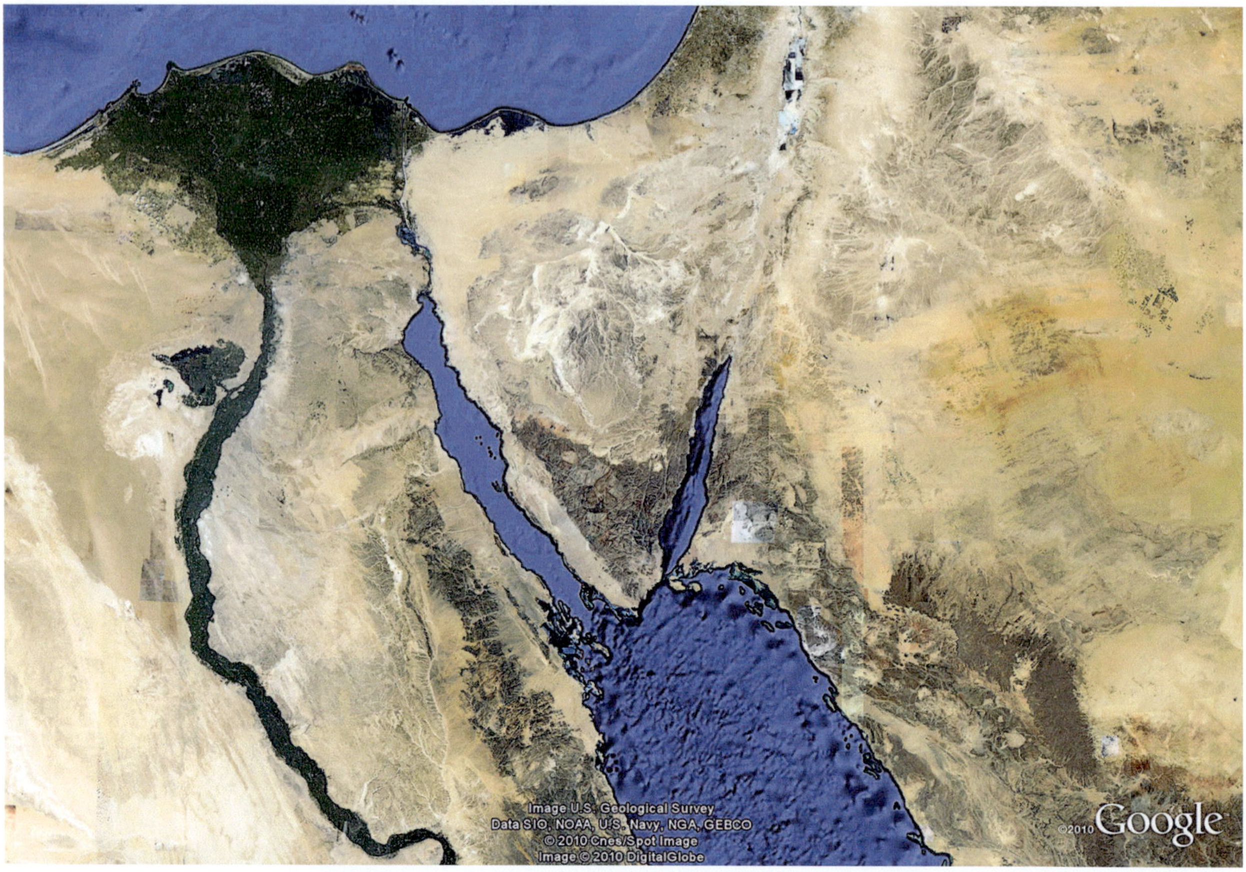
Image U.S. Geological Survey
Data SIO, NOAA, U.S. Navy, NGA, GEBCO
© 2010 Cnes/Spot Image
Image © 2010 DigitalGlobe
©2010 Google

2.11

Data SIO, NOAA, U.S. Navy, NGA, GEBCO
Data CSUMB SFML, CA OPC
Image © 2010 TerraMetrics
©2010 Google

long open fissures resulting from drifting apart of the pack, or high pressure ridges due to collision. If smaller pieces collide during the drift, they often change their form into typical "pancake" ice (Fig. 2.9). The pancakes start with a diameter of tens of centimeters, but through wind and wave action they aggregate with loose frazil crystals to increase in diameter, and raft with other pancakes to increase in thickness. In this manner the pancakes can rapidly grow to a few meters in diameter and up to a metre thick. Pancake-like ice patches with elevated outer rims can also form by collision of irregular fragments.

2.3 Structural Dominated Coastlines

In the following chapters we look at the large catalogue of "primary" coasts which are rather young in geologic terms and are shaped by terrestrial processes, including erosion, river/stream deposition, glaciers, volcanism, and tectonic movements. They have been submerged by the rising sea level after the termination of the last ice age. Along primary coasts, marine or littoral processes such as waves, currents or tides have not contributed significantly to the coastal landscapes or features. Let us start with a very important one: Coastlines dominated by tectonic movements of the Earth's crust such as faults, folds or joints. Faults and folds are examples of the basic features geologists observe and map to understand crustal deformation. Folds in rocks are like folds in clothing. Just as cloth pushed together from opposite sides bunches up in folds, layers of rock slowly compressed by forces in the crust can be pushed into folds. These forces can also cause a rock formation to break and slip on both sides of a fracture, forming a fault. Geologic folds and faults can range in size from centimeters to hundreds of kilometres.

The continental outlines have undergone major changes during Earth's history. About 300 million years ago, it is thought that all continents formed one huge landmass which geologists call the supercontinent Pangaea – the landmass that was the focus of Alfred Wegener's concept of continental drift. Pangaea started to breakup and drift apart about 200 million years ago, first into two landmasses, Laurasia near the North Pole and Gondwana near the South Pole, separated by the Tethys ocean. Subsequently, Laurasia and Gondwana split up into the landmasses which form our modern continents and plate tectonic drift has carried the continental fragments into their modern positions. Break up and movement

◄ Fig. 2.10 Continental drift and plate tectonics may form first order coastal features dominated by a structural origin. At the northern end of the Red Sea, a large graben structure formed by diverging of the African plate from the Arabian plate, spreads out into two smaller rifts flanking the Sinai peninsula. While the Gulf of Aquaba at right is very deep (more than 1000 m) and continues on land as the Dead Sea and Jordan graben, the Gulf of Suez is shallow. The Indian Ocean started to drown these grabens less than two million years ago. Width of image about 1000 km. (Photo credit: ©Google Earth 2010).

◄ Fig. 2.11 The San Andreas Fault strikes out into the sea north of San Francisco. (Photo credit: ©Google Earth 2010).

NEXT PAGES

► Fig. 2.12a,b Examples of hidden structures stripped by glacial erosion in fjord landscapes. The image below is from British Columbia (Canada, *54° 40′N* and *130° 30′W*): a conjugate system of joints was eroded by glacier flow during the ice ages and produced this pattern of fjords today. Width of image about 25 km. (Photo credit: ©Google Earth 2010).

► Fig. 2.13a,b The Belcher Islands in the Hudson Bay (Canada) at about *56° 10′N* and *79° 20′W* show these folding patterns, eroded and polished by several glaciations, again also in the form of "canale". Width of image about 220 km. (Photo credit: ©Google Earth 2010).

► Fig. 2.14 The "canale" coastline of former Yugoslavia in the Adriatic Sea, Mediterranean. (Photo credit: ©Google Earth 2010).

► Fig. 2.15 More or less parallel folds run out into the Indian Ocean in north-western Australia at about *16° 30′S* and *123° 40′E*. The parallel waterways between the island chains are called "canale" or "vallone" from Italian examples. Width of image is about 160 km. (Photo credit: ©Google Earth 2010).

► Fig. 2.16 Another example of a canale coast in the South China Sea at about *21°N* and *107° 30′E* near the northern part of Vietnam also exhibits old folds as a structural control, determined by past geological processes and partly drowned by the modern high sea-level. Width of image about 70 km. (Photo credit: ©Google Earth 2010).

► Fig. 2.17 More than 2 billion years old quartzite rocks in NW Australia (at about *15° 20′S* and *124° 30′E*) are deeply fractured by tectonic movements. Weathering and erosion has opened the joint pattern, and postglacial high sea-levels have flooded into small bays along these joints. Width of image about 15 km. (Photo credit: ©Google Earth 2010).

Image © 2010 TerraMetrics
Image © 2010 DigitalGlobe
Data Living Oceans Society
©2010 Google

2.12a

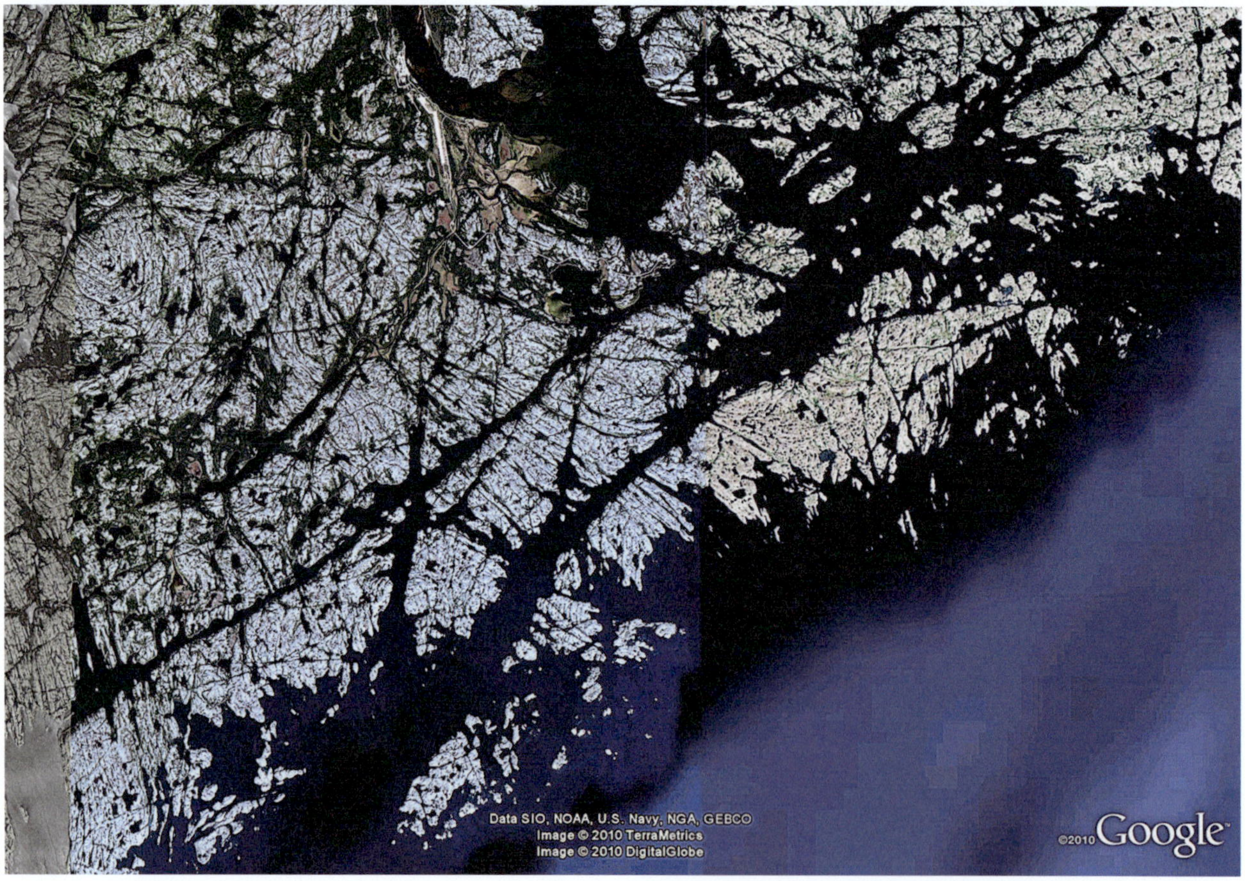
Data SIO, NOAA, U.S. Navy, NGA, GEBCO
Image © 2010 TerraMetrics
Image © 2010 DigitalGlobe
©2010 Google

2.12b

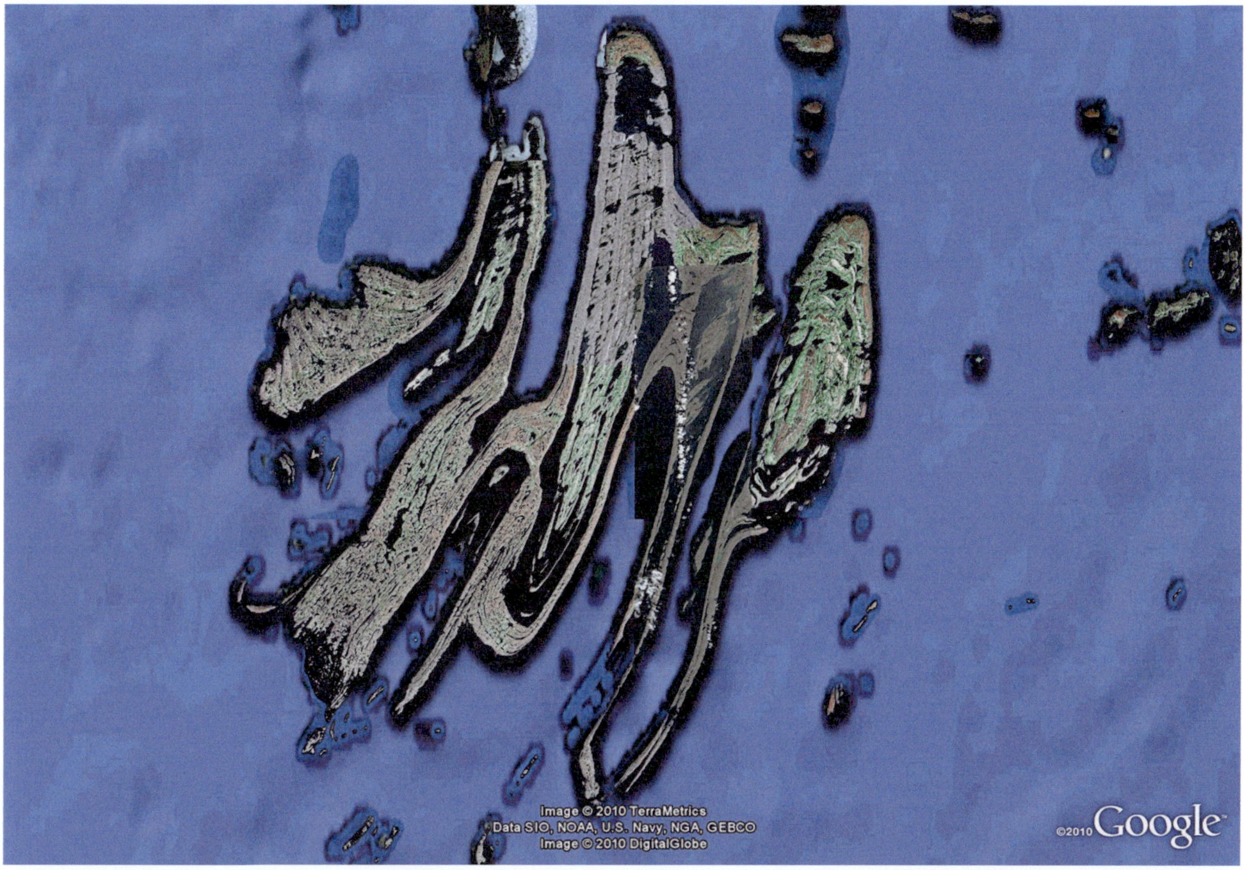
Image © 2010 TerraMetrics
Data SIO, NOAA, U.S. Navy, NGA, GEBCO
Image © 2010 DigitalGlobe
©2010 Google

2.13a

Image © 2010 TerraMetrics
Data SIO, NOAA, U.S. Navy, NGA, GEBCO
Image © 2010 DigitalGlobe
©2010 Google

2.13b

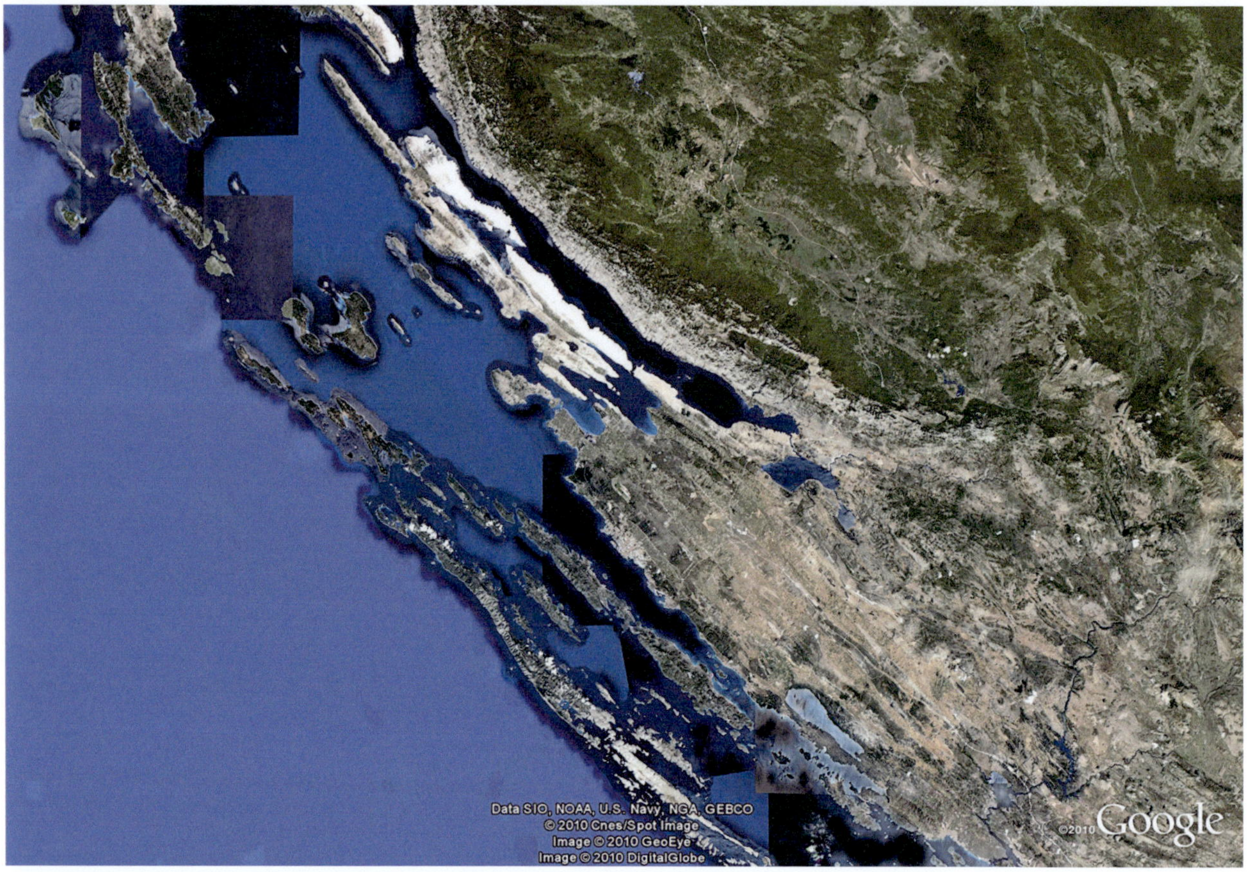
Data SIO, NOAA, U.S. Navy, NGA, GEBCO
© 2010 Cnes/Spot Image
Image © 2010 GeoEye
Image © 2010 DigitalGlobe
©2010 Google

2.14

© 2010 Cnes/Spot Image
Image © 2010 DigitalGlobe
Data SIO, NOAA, U.S. Navy, NGA, GEBCO
©2010 Google

2.15

Image © 2010 TerraMetrics
Image © 2010 GeoEye
©2010 Google

2.16

Image © 2010 DigitalGlobe
©2010 Google

2.17

of the plates continue today, as seen where the Arabian Peninsula is separating and moving apart from Africa giving birth to a new, linear ocean – the Red Sea (Fig. 2.10) which started to form about 2 million years ago.

The tectonic plates come together along convergent plate margins, move apart along divergent plate margins, and also slide past one another. A famous example of the latter is the San Andreas Fault (Fig. 2.11). The San Andreas Fault is a transform fault that separates the Pacific Plate from the North American Plate. San Francisco on the North American Plate is moving south while Los Angeles on the Pacific Plate is moving north at a speed of 5.5 cm per year. During this process, the plates grind past one another resulting in the build up of tension in the rocks. At times the plates get locked and when this locked section suddenly breaks, earthquakes occur.

Structures made by folding in former geological times may be exposed if glaciers have polished the landscape and unearthed these features from their cover of weathered rock and soil (Fig. 2.13). A special form of coastal landscape with elongate and often parallel islands chains can be found in areas where folded structures run parallel to each other. Anticlines (layered rocks that upfold into arches) can be emergent, whereas the synclines (those that downfold as in a trough) are drowned. From a bird's eye view a picture resembling parallel canals results, thus coastal landscapes dominated by those forms were given the name "canale" (or "vallone" = valley) from the Italian terms (Figs. 2.14 to 2.16). This type of coastline is a real challenge for marine navigators, for although the sea in these narrow channels is mostly smooth and without large wave action, strong tidal currents can develop in the narrow passages if the tidal range is high.

Joints are fractures in which there has been no movement along the fracture. They can be formed in a number of ways, including the cooling and shrinkage of igneous (volcanic) rocks, mechanical stress during uplift or subsidence or the release of confining pressure associated with physical weathering. Joints are usually only the beginning of a series of changes that significantly alter geologic formations as they age. They provide channels through which water (either from the sea or as rain) and air can reach deep into the formation and speed the weathering and internal weakening of the rock. Coastlines dominated by exposed joints often exhibit a criss-cross pattern of straight lineaments from a bird's eye perspective. In higher latitudes, grinding glaciers may erode joints in older rocks and thus shape the fjord landscape we see today (Fig. 2.17).

Marine cliffs are prominent features of the coastal scenery along structural dominated coastlines and contribute tremendously to the aesthetic perception of coastal landscapes. The type of scenery that develops on cliff coasts is the product of a number of factors such as the morphology of the hinterland, present and past climates, wave and tidal environments, changes in relative sea level and the structure and lithology of the rocks. Small bays, narrow inlets (also called geos), caves, arches and stacks are usually the result of wave erosion impacting cliff coastlines particularly along joint and fault planes, or in faulted and structural weakened rock formations.

2.4 Volcanic Coasts

Volcanoes are fascinating because they are windows through which we can see into Earth's deep interior to understand the processes of plate tectonics that have modified Earth's oceanic and continental crust. The shape of a volcano varies with the properties of the magma, especially its chemical composition and gas content and the environmental conditions under which the lava erupts, such as on land or under the sea. Plate tectonics can explain essentially all major features in the global pattern of volcanism and the locations of the world's active volcanoes that occur on land or above the ocean's surface. It is in the vicinity of plate margins where volcanoes and oceans are intimately interconnected (Figs. 2.18, 2.19 and 2.20): About 80 percent are found at boundaries where plates converge (as for example around the subduction zones of the Pacific Ocean in the "Ring of Fire"), 5 percent where plates separate, and the remaining few within the plate interiors, at so called Hot-Spots. These volcanoes occur where a plate moves over a very hot area in the asthenosphere and magma rises to the surface – sometimes these are of extraordinary dimensions as for example in the Hawaii Islands where Mauna Loa on "Big Island" is the world's largest

Fig. 2.18 A group of young volcanoes dominates the coastline configuration in the Aleutian Islands of Alaska at about *53°N* and *170°W* with a width of the scene of about 40 km. (Photo credit: ©Google Earth 2010).

active volcano in terms of volume and area covered. Mauna Loa is projecting 4100 m above sea level, but the flanks of Mauna Loa sit on sea floor that is about 5000 m deep. From its base below sea level to its summit the "height" of this volcano relative to the sea floor is 9170 m – Mauna Loa is taller than Mount Everest! The trace of a hot spot such as this appears as an island chain: The hot spot remains constant and as the overlying plate continues to move over it thus the older volcanoes move away from the volcanic source and become extinct with successively newer ones forming over top of the hot spot. The hot spot that is currently under the island of Hawaii has created a chain of islands and seamounts that extend from Hawaii to the Aleutian trench.

The plate tectonic model also explains the presence of volcanic island arcs parallel to oceanic trenches along subduction zones. When a plate undergoes subduction it encounters hotter conditions deep in the Earth's crust. As a result, water is driven out of the oceanic crust which has the effect of lowering the melting temperature of rocks. Some of the magma generated by this melting rises to the surface and forms the volcanoes of the island arc. The volcanoes of island arcs and active plate margins tend to be explosive as a result of water gaining access to the rising magma through fissures and being vaporized into superheated steam. The resulting volcanic debris forms tephra cones which survive only briefly in geological terms but which may contribute to coastal configurations (Fig. 2.24).

Calderas result when a violent eruption empties a volcano's magma chamber, which then cannot support the overlying rock. It collapses and leaves behind a large, steep-sided basin. Caldera collapse of volcanic edifices may result in magnificent harbours and embayments (Fig. 2.21 to 2.23).

Santorini eruption (~1628 BC) and the legend of Atlantis

Volcanic eruptions have a prominent place in human history and mythology and ancient philosophers were awed by volcanoes and their fearsome eruptions of molten rock. In their efforts to

Fig. 2.19 Lava flows form all coasts of Kaimeni Island in the Caldera of Santorini, Greece. (Photo credit: ©Google Earth 2010).

explain volcanoes, they spun myths and legends about a hot, hellish underworld below Earth's surface. The word "volcano" comes from the small island of Vulcano in the Mediterranean Sea off Sicily. Centuries ago, people living in this area believed that Vulcano was the chimney of the forge of Vulcan – the blacksmith of the Roman gods. They thought that the hot lava fragments and clouds of dust erupting form Vulcano came from Vulcan's forge as he beat out thunderbolts for Jupiter, king of the gods, and weapons for Mars, the god of war.

Fig. 2.20 Lava flows into the sea on the Galapagos Islands, Ecuador. (Photo credit: ©Google Earth 2010).

The following account may show how volcanic coastlines have inspired man's fantasy and imagination: A famous caldera is the one of Santorini Island in Greece, also called Thera (Fig. 2.23). The Minoan eruption of Thera, called the Santorini eruption, was a major catastrophic volcanic explosion and collapse during the Bronze Age around 1628 BC. The eruption was one of the largest volcanic events in recorded history and inspired the myth of Atlantis, first recorded by Plato, the Greek philosopher. In the modern era, geologic and archaeological investigations hint at an intriguing possibility – that the myth of Atlantis may be related to this catastrophic eruption in the Aegean Sea, which generated a flooded caldera and destroyed an advanced Minoan civilisation living on the island group of Santorini. The Greek philosopher Plato (427–347 BC) describes in his dialogs "Critias and Timaeus" the disappearance of Atlantis, a circular island with circular canals populated by talented people of culture and wealth. Plato's account was originally derived from Solon (640–560 BC), the great sage and lawgiver from Athens. While visiting the town of Sais on the Nile delta, Solon was told by Egyptian priests of the disappearance of a great island empire. The story was passed to Plato from Critias, through his great grandfather who had discussed the story with Solon. All subsequent writings and speculations about Atlantis are rooted in Plato's dialogs. In Timaeus, Plato quotes Critias' account of the legend, as told to Solon by

NEXT PAGES

►Fig. 2.21 A volcanic collapse into the sea (most probably associated with a tsunami) formed the NW coast of Hierro Island in the Canaries, Spain, at *27° 45′N and 18°W*. The island is approximately 23 km long. (Photo credit: ©Google Earth 2010).

►Fig. 2.22 A volcanic coastline of Japan, showing a drowned caldera with a diameter of about 7 km at *44° 35′N and 147°E*. (Photo credit: ©Google Earth 2010).

►Fig. 2.23 The drowned caldera of Santorini, Aegean Sea, Greece, with up to 11 km diameter at *36° 24′N and 25° 24′E* from a collapse about 3600 years ago. In the centre is the younger island of Nea Kaimeni, whose lava flows are shown in Fig. 2.19. (Photo credit: ©Google Earth 2010).

►Fig. 2.24 A nearly drowned crater of a tephra volcano with a diameter of 0.5 km in the Galapagos group of Ecuador at *0° 23′S and 91°W*. (Photo credit: ©Google Earth 2010).

Data SIO, NOAA, U.S. Navy, NGA, GEBCO
Image © 2010 GRAFCAN
©2010 Google

2.21

Data © 2010 MIRC/JHA
Image © 2010 DigitalGlobe
©2010 Google

2.22

Image © 2010 DigitalGlobe
Data SIO, NOAA, U.S. Navy, NGA, GEBCO
©2010 Google

2.23

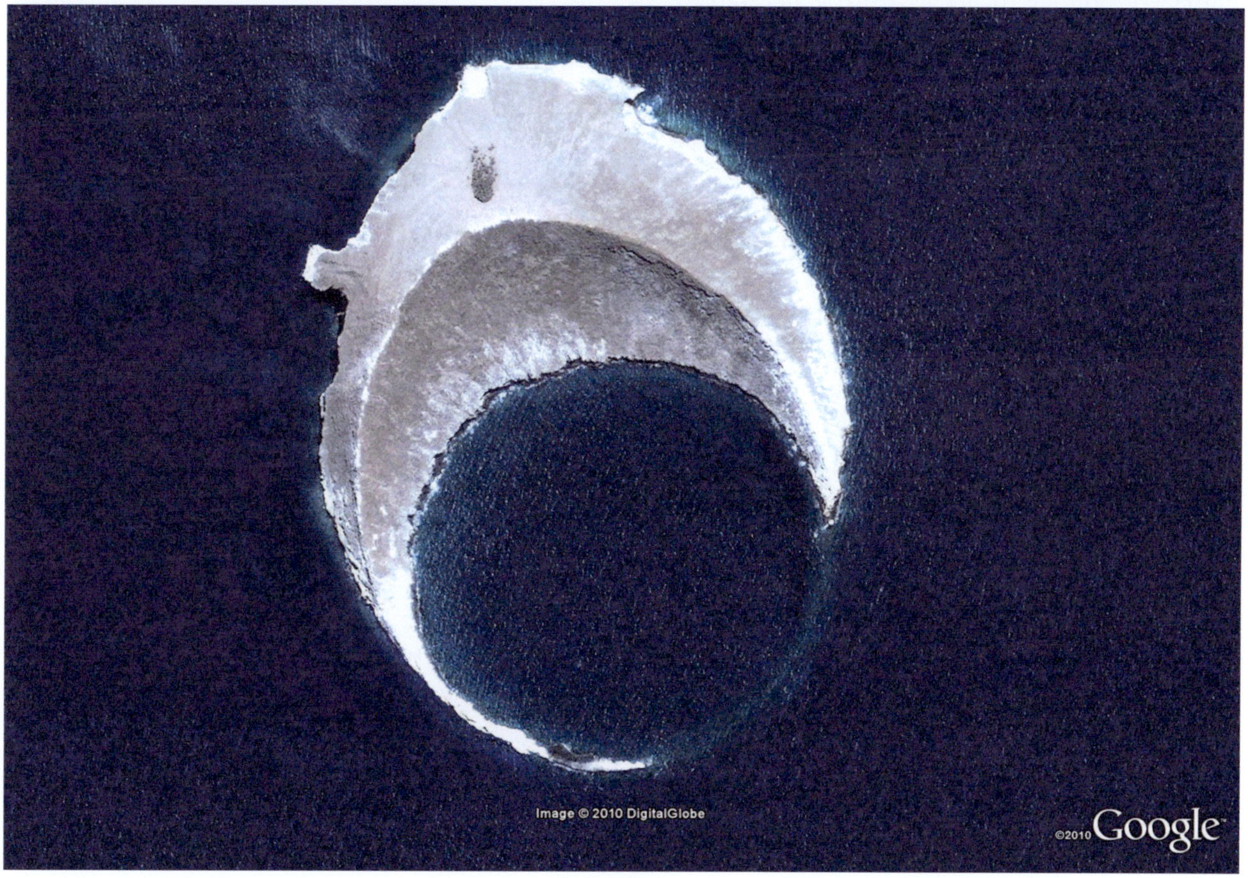
Image © 2010 DigitalGlobe
©2010 Google

2.24

one of the Egyptian priests: *"Now in this island of Atlantis there was a great and wonderful empire which had rule over the whole island and several others, and over parts of the continent. But, there occurred violent earthquakes and floods, and in a single day and night of misfortune the island of Atlantis disappeared in the depths of the sea."*

The geologic record at Santorini reveals a long history of volcanic activity, consistent with its subduction-zone setting. The archaeological record indicates that Santorini has been inhabited by civilizations going back to the 17th century BC, contemporaneous with the most recent eruptive events. Archeological excavations at Akrotiri in southern Thera have revealed Bronze Age ruins of a particularly large and vibrant city, with well-preserved frescos and paintings in 2-storey houses, together with numerous artefacts. The artefacts indicate that the island of Thera was colonized by the Minoans, a Bronze Age civilization named after the legendary King Minos of Crete. Thera appears to have had a thriving Minoan economy provided by intensive trade throughout the eastern Mediterranean. Today, the remains of this flourishing community lie buried under a thick blanket of pumice (of up to 10 m) generated by a massive Late Bronze Age eruption. The exact date of the eruption remains somewhat controversial, although most radiometric studies show that it falls between 1615–1645 BC, consistent with a pronounced acid-ice layer from the Greenland cores, dated at 1636 BC. The event was of epic dimensions and there is only one eruption in human history believed to have been larger: an 1815 explosion of Tambora, in Indonesia, which released 100 cubic kilometers of volcanic products (lava and tephra).

References

Kelletat D (1999) Physische Geographie der Meere und Küsten. Teubner Studienbücher Geographie, 2 ed., Stuttgart

Finkl CW (2004) Coastal classification: Systematic approaches to consider in the development of a comprehensive system. Journal of Coastal Research, 20(1), 166–213.

Stone GS (2003) Ice Island: The Expedition to Antarctica's Largest Iceberg. Bunker Hill Publishing. Charlestown, Massachusetts, U.S.A.

Woodroffe CD (2003) Coasts: Form, Process and Evolution. Cambridge University Press, Cambridge.

3

Coastlines Dominated by Ingression of the Sea into older Terrestrial Landforms

ABSTRACT At shorelines, we can observe the constant motion of ocean waters and their effects. Sea-level changes, waves, longshore currents, and tidal currents interact with the rocks and tectonics of the coast to shape shorelines into a multitude of forms. Coastal scientists refer to primary coasts as a category of coastal landscapes which show well-preserved features of former terrestrial morphologies. Ingression coastlines are simply drowned coastal landscapes as ingression refers to the advance of the sea into existing terrestrial topography, like the drowning of a river valley by the Holocene sea level rise. They occur in all latitudes, and in fact all morphological features of continental landscapes may appear as drowned coastal forms and landscapes. The rich diversity of ingression coastlines which owe their preservation to sheltered positions along the shoreline and to the short duration of the recent high sea level (6000 to 7000 years) is presented in 35 figures.

3.1 Ingression in Rocky Glacial Landscapes

The landscapes of many continents have been sculpted by glaciers now melted away. In mountainous regions, glaciers have eroded steep-walled valleys, scraped bedrock surfaces, and plucked huge blocks from their rocky floors. During the ice ages, glaciers pushed across entire continents, carving off far more topography than rivers and wind. A small valley glacier only a few hundred metres wide can tear up and crush millions of tons of bedrock in a single year. It carves a series of erosional forms as it flows from its origin to its lower edge. At the head of the glacial valley, the plucking and tearing action of the ice tends to carve out an amphitheatre-like hollow called a cirque. As a valley glacier moves down slope it deepens the existing river valley or excavates a new one, creating a characteristic U-shaped cross section. At coastlines, valley glaciers may erode their floors far deeper than sea level and when the ice retreats, these steep-walled U-valleys are flooded with seawater. These former glaciated valleys now flooded by arms of the sea are called fjords (Figs. 3.1–3.3).

Often the patterns of fjords are adapted to old structural elements and the main faults and joints of the surrounding country rock (Fig. 3.4); in some cases the intersections of one fjord with another may be at a sharp angle. Fjords create the spectacular rugged scenery for which the coastlines of former Pleistocene glaciated regions are renowned – from Alaska to British Columbia, Kamchatka to northern Japan, Greenland, Scotland and Norway in the Northern Hemisphere. And in the Southern Hemisphere fjord landscapes can be seen in Chile, Tierra del Fuego (Argentina), the islands of the southern Atlantic Ocean and along the west coast

Fig. 3.1 Glacier erosion will carve and transform valley transects to wide "U"-forms. With glacier recession the ocean invades the valleys, resulting in a "fjord". This picture is from south-eastern Greenland at about *62°N and 42° 20′W* with a width of about 80 km. (Photo credit: ©Google Earth 2010).

© 2010 TerraMetrics
Image IBCAO
©2010 Google

Fig. 3.2 Fjord coast along the west coast of southern Chile at about *44° 30´S* and *74°W.* Width of the scene is about 110 km. (Photo credit: ©Google Earth 2010).

◄ Fig. 3.3 Milford Sound, a fjord in southern New Zealand. The mountains in the background are up to 2000 m high. Water depth is nearly 1000 m. (Photo credit: D. Kelletat).

► Fig. 3.4 This fjord and skerry coast in southern Chile at *48°S* and *75°W* shows clear signatures of an old folding and jointing pattern in a scene about 30 km wide. (Photo credit: ©Google Earth 2010).

► Fig. 3.5 Drowned cirques, the upper armchair-shaped hollows with a steep back wall, of former glaciers in the Aleutian island chain of Alaska at about *55°N* and *160°W*, with a N-S width of the scene of 25 km. (Photo credit: ©Google Earth 2010).

Data SIO, NOAA, U.S. Navy, NGA, GEBCO
Image © 2010 TerraMetrics
©2010 Google

3.4

Data SIO, NOAA, U.S. Navy, NGA, GEBCO
Image © 2010 DigitalGlobe
Image © 2010 TerraMetrics
©2010 Google

3.5

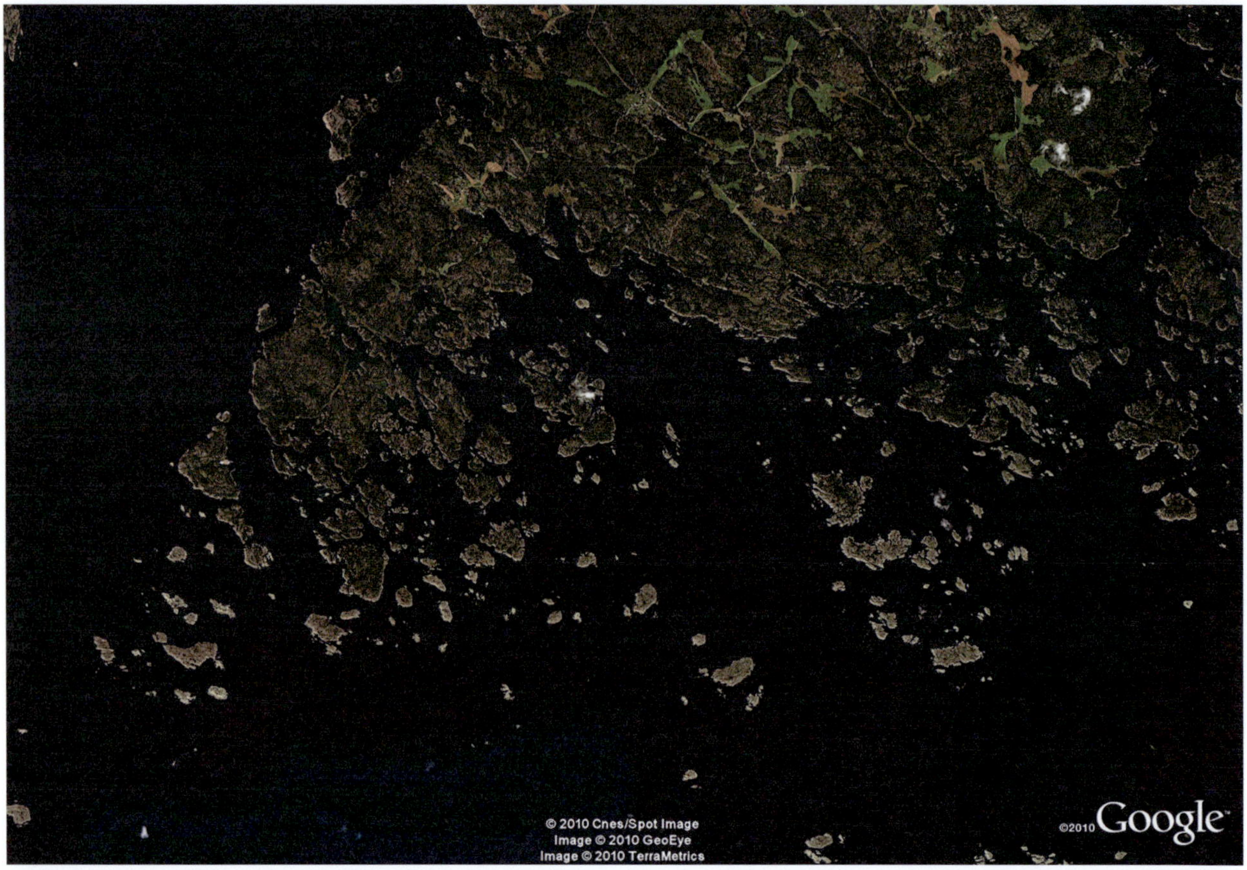
© 2010 Cnes/Spot Image
Image © 2010 GeoEye
Image © 2010 TerraMetrics
©2010 Google

3.6a

© 2011 Cnes/Spot Image
Data SIO, NOAA, U.S. Navy, NGA, GEBCO
©2010 Google

3.6b

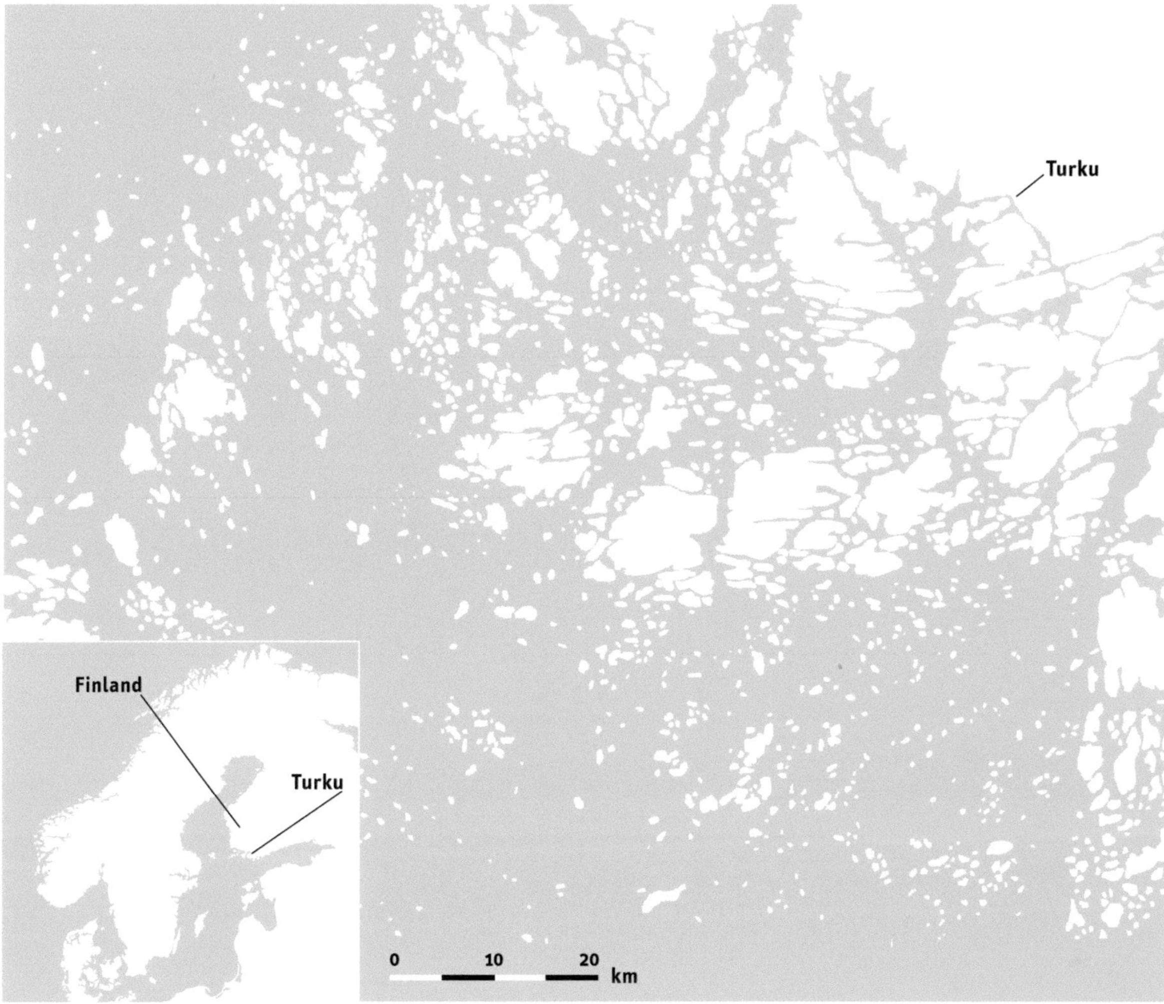

Fig. 3.7 Skerries of the Åland Archipelago west of southern Finland. (Image credit: D. Kelletat)

of the South Island of New Zealand. Fjords are ingressional coastal landforms and the long profile of many fjords, including alternating basins and steps, is that of the former glaciated valleys. Over their longitudinal profile, fjords may reach great depths (even over 1000 m!), however, the floor normally rises steeply to create a rock threshold towards the mouth and water depths decrease. In some regions even the upper parts of glaciated valleys or cirques may be drowned. They result in semi-circular bays surrounded by steep mountains. Good examples can be found in the Aleutian island chain (Fig. 3.6) as well as at the Lofoten Islands in northern Norway.

◂ Fig. 3.6a,b A skerry coastline from the Åland Archipelago of Finland, Baltic Sea, at *60°N* and *20°25′E*, 12 km wide. (Photo credit: ©Google Earth 2010).

Pleistocene continental ice caps may have also carved out wide depressions along relatively low-lying coasts which by transgression of the post glacial sea level have been transformed into fjärds – rocky inlets of the sea. A typical feature of fjärds are small rocky islands called skerries and partly drowned roches moutonnées (Figs. 3.4–3.8). A roche moutonnée ("sheep rock" for its resemblance to a sheep's back) is a small bedrock hill, smoothed by the ice on the upstream side and plucked to a rough face on the downstream side due to the moving ice pulling fragments from joints and cracks. Skerries are also typical in fjord landscapes, but mostly beside their mouths to the open ocean.

Data SIO, NOAA, U.S. Navy, NGA, GEBCO
Image © 2010 TerraMetrics
©2010 Google
3.8

Image © 2010 TerraMetrics
Data SIO, NOAA, U.S. Navy, NGA, GEBCO
Image © 2010 DigitalGlobe
©2010 Google
3.9

►Fig. 3.10 Drumlins partly drowned near Yakutat in southern Alaska. (Photo credit: D. Kelletat).

◄Fig. 3.8 A coastline with drowned drumlins in northern Massachusetts, USA, at *42° 17′N* and *70° 54′W*, with a width of 12 km. (Photo credit: ©Google Earth 2010).

◄Fig. 3.9 Drowned drumlins at the west coast of Ireland at *53° 50′N* and *9° 40′W*. Width of scene 10 km. (Photo credit: ©Google Earth 2010).

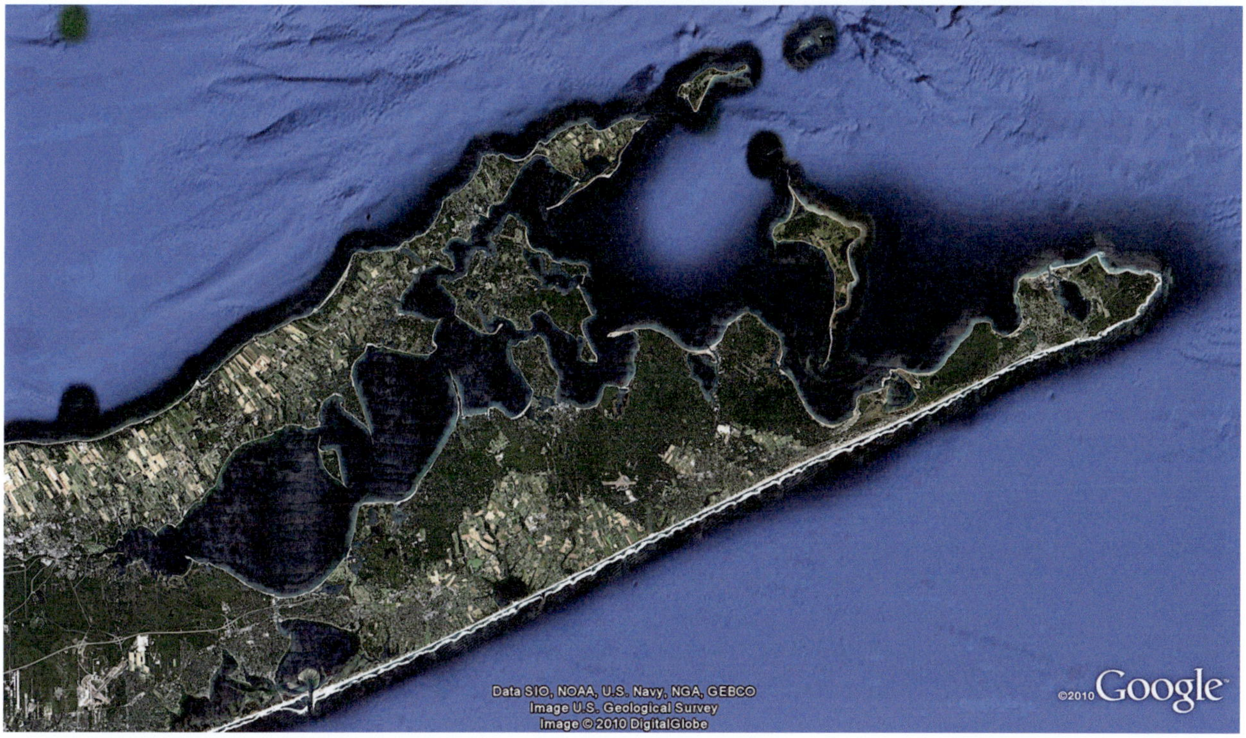

Fig. 3.11 The outer contours of Long island east of New York (USA) are formed by long terminal moraines about 25 km apart at *41°N* and *72° 10′W*. (Photo credit: ©Google Earth 2010).

3.2 Ingression in Sedimentary Glacial Landscapes

Glacial erosion creates enormous amounts of debris and ice transports huge volumes of sediment depositing it at the edges of glaciers where it may be carried away by meltwater streams. Glacial processes affect the water discharge and sediment loads of major river systems, the erosion and sedimentation of coastal areas, and the quantity of sediment delivered to the oceans. When glacial ice melts, it deposits a poorly sorted, heterogeneous load of boulders, pebbles, sand, and clay, which puzzled the early geologists who called it drift because it seemed to have drifted in somehow from other areas. Today we call this material till or moraine. Meltwater streams flowing in tunnels within and beneath the ice and in streams at the ice front may pick up, transport, and deposit some of the material either under the ice as eskers which are long, narrow, winding ridges of sand and gravel, or as whale-like asymmetrical but streamlined hills of till and bedrock termed drumlins. The postglacial sea level rise flooded these landforms of the last ice age along the coastlines of higher latitudes, as you can see in Denmark or along the west coast of Ireland (Figs. 3.9–3.11). At some places terminal moraines marking the maximum advance of past glaciers are preserved as elongated shore parallel islands or island chains in shallow water. Good examples are to be seen east of New York (Fig. 3.12).

3.3 Ingression into Fluvial Landscapes

Streams of all sizes – from tiny rills to majestic rivers – are major geological agents of change and prominent agents of the erosion of the. Most streams have eroded well-defined channels in bedrock or in unconsolidated sediment and incised valleys which allow water to flow over great distances. In contrast to glacial valleys and their typical U-shape, the cross-sectional profile of many river valleys is V-shaped, but many other valleys have a broad, low profile. If you follow a stream from its mouth to its head, you will see that it steadily divides into smaller and smaller tributaries, forming drainage networks that show a characteristic branching pattern. Perhaps the most familiar irregular branching pattern resembles a tree called dendritic drainage. This fairly random drainage pattern is typical of landscapes where the bedrock is uniform, such as horizontally bedded sedimentary rocks or massive volcanic or metamorphic rocks. Other drainage patterns are rectangular, trellis, and radial, controlled by systems of joints or faulting and folding in the bedrock.

Global sea level fluctuated quite wildly over geologic time, especially during the ice ages of the Pleistocene epoch. Thus, not surprisingly river valleys have been incised below present sea level during sea level lowstands and the postglacial transgression drowned this fluvial relief incised in the shelf areas and the edges of the continents (Figs. 3.13–3.23).

We can distinguish between complex fluvial landscapes or drainage patterns, where ingression is the most prominent process and coastal landscapes in which just one prominent valley has been flooded in its lower course. To these the Spanish term "ria" (pl. rias) is applied derived from large inlets on the coasts of Galicia. In Brittany and Wales they are known as abers, along the higher limestone coasts in the Mediterranean as calas or calanques, while those on arid coasts are termed sharms or sherms (Fig. 3.17). Where the coastal relief is rather steep and therefore rias are short and exposed to wave action, the former fluvial forms are not only drowned, but transformed into embayments or indented coastlines dominated by selective abrasion. For rias in humid climates,

►Fig. 3.12 Drowned valleys or typical rias along a 7 km long limestone island of former Yugoslavia at *43°N* and *16°22′E*. (Photo credit: ©Google Earth 2010).

►Fig. 3.13 Rias in the eastern part of Papua New Guinea at *9°06′S* and *149°17′E*. Width of scene is about 20 km. (Photo credit: ©Google Earth 2010).

NEXT PAGES

►Fig. 3.14 The Banks Peninsula near Christchurch on the east coast of New Zealand′s south island at *43°45′S and 172°50′E* is an old volcano with many rias around it. Width of scene 60 km. (Photo credit: ©Google Earth 2010).

►Fig. 3.15 A fluvial form with a narrow cut through an uplifted coral reef and wider basins inland have been drowned by the post-glacial sea-level rise in northern Haiti at *19°40′N* and *71°50′W*. Width of scene is 21 km. (Photo credit: ©Google Earth 2010).

►Fig. 3.16 Rias in desert landscapes, where there are no longer active valleys, are also called "sherms". Example from the Red Sea coast of Egypt, 10 km wide, at *21°48′N* and *36°52′E*. (Photo credit: ©Google Earth 2010).

►Fig. 3.17 A "hierarchy" of ria forms of different size on the east coast of the USA at *38°45′N* and *76°9′W*. Width of image is about 22km. (Photo credit: ©Google Earth 2010).

►Fig. 3.18 Effective drowning of a fluvial landscape by sea-level rise and subsidence in the region of Marlborough Sounds, NW part of New Zealand's south island.

►Fig. 3.19 A good example of a drowned landscape dominated by valleys, now rias. (Northern part of Oman, Arabian peninsula at *26°10′N* and *56°24′E*). Scene is 35 km wide. (Photo credit: ©Google Earth 2010).

►Fig. 3.20 A ria coast from southern Japan (Nagasaki) at approximately *32°54′N* and *129° E*. Scene width is 24 km. (Photo credit: ©Google Earth 2010).

►Fig. 3.21 The ria system of Sydney Harbour, New South Wales, Australia, at *33°50′S* and *151°22′E*. Scene is about 25 km wide. (Photo credit: ©Google Earth 2010).

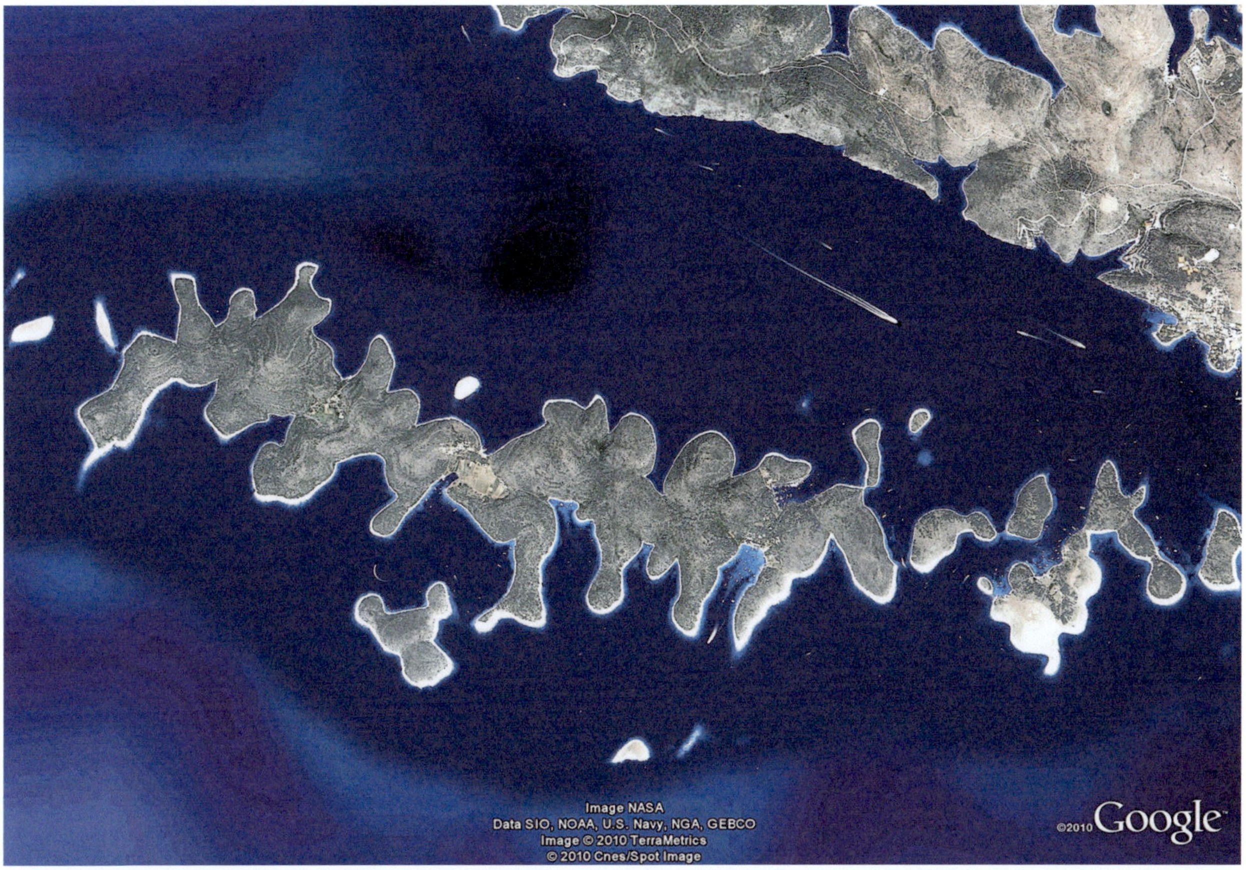
Image NASA
Data SIO, NOAA, U.S. Navy, NGA, GEBCO
Image © 2010 TerraMetrics
© 2010 Cnes/Spot Image
©2010 Google

3.12

Image © 2010 DigitalGlobe
Data SIO, NOAA, U.S. Navy, NGA, GEBCO
©2010 Google

3.13

Image © 2010 TerraMetrics
Data SIO, NOAA, U.S. Navy, NGA, GEBCO
Image © 2010 DigitalGlobe
Image © 2010 GeoEye
©2010 Google

3.14

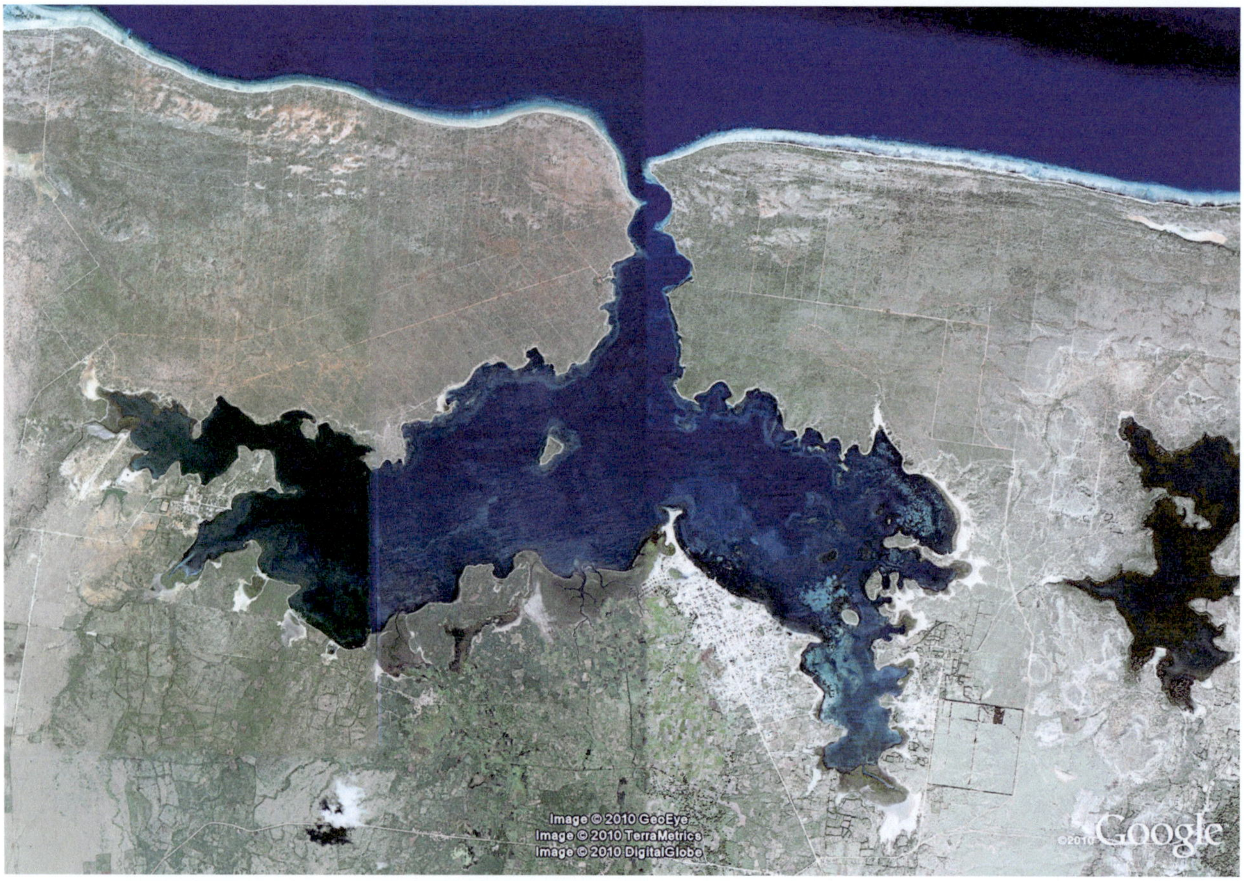
Image © 2010 GeoEye
Image © 2010 TerraMetrics
Image © 2010 DigitalGlobe
©2010 Google

3.15

Image © 2010 DigitalGlobe
©2010 Google

3.16

Image © 2010 TerraMetrics
Image U.S. Geological Survey
Data SIO, NOAA, U.S. Navy, NGA, GEBCO
©2010 Google

3.17

Image © 2011 DigitalGlobe
Image © 2011 GeoEye
Data SIO, NOAA, U.S. Navy, NGA, GEBCO
Image © 2011 TerraMetrics
©2010 Google

3.18

Data SIO, NOAA, U.S. Navy, NGA, GEBCO
Image © 2010 GeoEye
©2010 Google

3.19

Image © 2010 GeoEye
Image © 2010 TerraMetrics
©2010 Google

3.20

Data SIO, NOAA, U.S. Navy, NGA, GEBCO
Image © 2010 Sinclair Knight Merz
©2010 Google

3.21

Image © 2010 DigitalGlobe
©2010 Google

Image © 2010 TerraMetrics
©2010 Google

◀ Fig. 3.22 Drowned valleys which are closed by barriers forming lagoons perpendicular to the main coastal contour are called "liman" according to examples from the Black Sea. This image is from Martha's Vinyard island, Massachusetts, USA, at *41° 21′N* and *70° 36′W*, width of the scene of about 12 km. (Photo credit: ©Google Earth 2010).

◀ Fig. 3.23 Different types of older terrestrial relief forms have been drowned by the postglacial sea-level rise. In this example from the west coast of Andros Island, northern Bahamas (*24° 35′N* and *78° 10′W* with a width of about 20 km) numerous shallow sink holes in a limestone terrain are subjected to this flooding, forming a unique coastal landscape. (Photo credit: ©Google Earth 2010).

with high rainfall and therefore high river discharge, for example in the monsoon regions in the tropics, transformation of the forms by sedimentation is much greater than in dry and rocky landscapes. In these areas some ria forms have been unchanged for millennia (Fig. 3.13). Most rias are estuaries in the sense that inflowing rivers provide freshwater that encounters and mixes with seawater that is moving in and out of the ria systems following tidal movements. These brackish water conditions may be home to a unique fauna and flora of foraminifera, diatoms, ostracods and higher organisms.

3.4 Ingression into Karst Landforms

A certain group of rocks is easily affected by solution if water is present through a process called karstification. It results in forms of variable scale, from tiny rills to extended landscapes. Karst weathering dominates some large regions of our planet such as on the larger islands of the inner Caribbean (Cuba, Haiti), or in southern China and Vietnam.

Fig. 3.24 Detail of partly drowned shallow sinkholes from Andros Island, Bahamas. (Photo credit: D. Kelletat).

Image © 2010 European Space Imaging
Data SIO, NOAA, U.S. Navy, NGA, GEBCO
Image © 2010 DigitalGlobe
©2010 Google

3.25

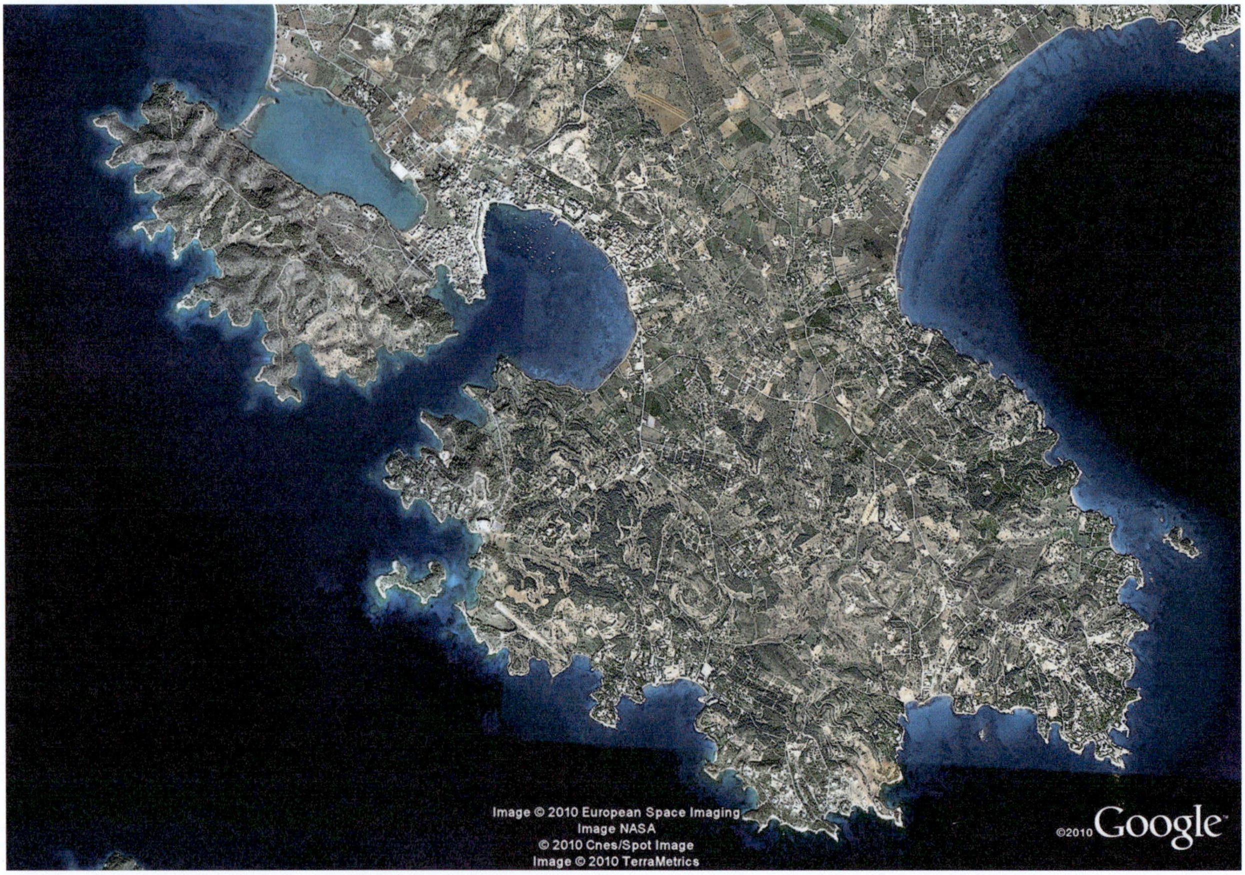
Image © 2010 European Space Imaging
Image NASA
© 2010 Cnes/Spot Image
Image © 2010 TerraMetrics
©2010 Google

3.26

◄ Fig. 3.25 Two nearly drowned poljes from Lesbos Island, eastern Aegean Sea, Greece, at *39° 09′N* and *26° 13′E*. The island is about 65 km wide. (Photo credit: ©Google Earth 2010).

◄ Fig. 3.26 Partly drowned poljes on the southern Argolid peninsula, Peloponnesus, Greece. (Photo credit: ©Google Earth 2010).

► Fig. 3.27 Examples from the cockpit and tower karst in the "Rock islands" of Palau, Micronesia. (Photo credit: D. Kelletat).

Fig. 3.28 Drowned tower karst with numerous small but very steep islands in the South China Sea at about *20° 46′N* and *107 04′E*. The scene is about 30 km wide. (Photo credit: ©Google Earth 2010).

Fig. 3.29a,b Coastal karst towers up to 250 m high in southern Thailand's Phang Nga Bay. (Photo credit: A. Scheffers).

Over long geologic times all types of rock may undergo solution, but only in a few cases solution dominates all other processes resulting in typical landscape features. These rocks are Halite (rock salt), Gypsum, and Limestone (calcium carbonate), and to a lesser extent, Dolomite. Over the same time unit, the amount of solution of these different rocks is equal to 1 time (salt): 100 times (gypsum): 10.000 times (limestone). In very humid areas solution is much faster than in arid (desert) climate zones.

The solution of these rock types results in distinctive and often spectacular landforms: sinkholes, cockpits and poljes are evident as depression forms in the landscape (Figs. 3.24–3.27); remnants of the relief that have resisted solution are mogotes, cones and karst towers (Figs. 3.28–3.30). Cockpit karst is characteristic of some limestone islands of the Caribbean (Cuba, Hispaniola or Guadeloupe). Tower karst is typical along the coastlines of southern China, Vietnam or Thailand with a humid and per-humid climate, whereas flooded poljes can be found particularly in the eastern Mediterranean (Figs. 3.26 and 3.27). Poljes are kilometre wide and long low lying fields. They occur where tectonic depressions have developed in limestone areas that are subsequently enlarged by solution along the margins. The flooding of karst depressions usually commences by infiltration of the sea through caves and joints enlarged by solution. Inland karst lakes near the coast may thus contain salt water and may have high concentrations from evaporation in warm latitudes. Marine organisms like jellyfish may be found in some lakes with underground connection to the sea.

3.5 Ingression into Eolian Landforms

The ancient Greeks called the god of winds Aeolus, and geologists today use the term eolian for the geological processes controlled by wind. Eolian processes shape the surface of the land, particularly in desert environments, where strong winds can howl

Fig. 3.30 Partly drowned longitudinal dunes from NW Australia at *17° 18′S* and *123° 21′E*. On the right side of the image a tidal flat with mangrove fringes can be seen. Width of scene is 9 km. (Photo credit: ©Google Earth 2010).

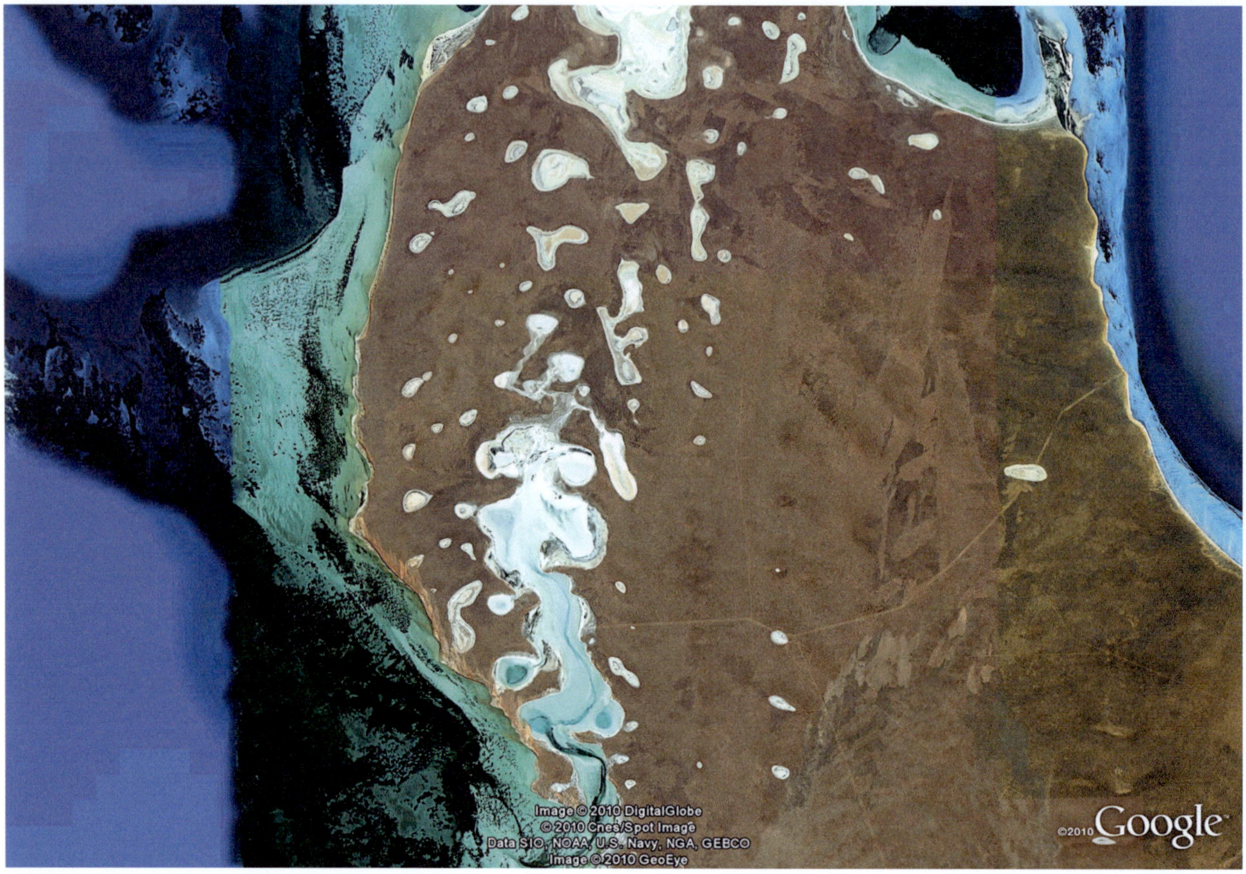

Fig. 3.31 Partly drowned deflation depressions at *25°46′S* and *113°29′E* in Shark Bay, Western Australia. Width of scene is nearly 40 km. (Photo credit: ©Google Earth 2010).

for days and create landforms through erosive and depositional processes. Wind can move enormous quantities of sand, silt, and dust over large regions of the continents and oceans and is much like water or ice in its ability to erode, transport, and deposit sediment in arid environments where not much moisture is present in the soil. As particles of dust, silt, and sand become loose and dry, winds can lift and carry them away, gradually eroding the ground surface in a process called deflation. Deflation scoops out shallow depressions on dry plains and deserts or on temporarily dried-up river floodplains and lake beds. Large deflation depressions form extended shallow bays along the coastline of Shark Bay, Western Australia (Fig. 3.32).

When the wind dies down, it can no longer transport the sediments it carries and deposits the coarser material such as sand in dune forms of various shapes and dimensions. The most dramatic sedimentary accumulations in a desert landscape and along sandy beaches all over the world are sand dunes. As most Holocene dunes are unconsolidated, they are prone to erosion by the sea through sea level rise or storm inundations. In general, they are not preserved as coastal landforms along submerging coasts, except in very sheltered places. Fig. 3.31 shows one of the rare examples preserved. Older dunes of the Pleistocene epoch often become partly cemented (lithified) by secondary internal precipitation of carbonates from percolating groundwater. Geologists call them eolianites. Eolianite dune formations may have been emplaced on a coast during oscillations of sea level over the Pleistocene ice ages, such as along the Nepean Peninsula in Victoria (Australia) where they extend more than 140 m below and 60 m above present sea level (Bird, 1993).

3.6 Permafrost Coastlines with Ingression

Perennially frozen ground, or permafrost, today covers as much as 25 percent of Earth's total land area and is common in higher latitudes in countries such as Alaska, Canada, Greenland, Spitsbergen or Siberia. In these regions, the repeated freezing and thawing of deep frost on the ground leads to some unusual landforms. Common features are depressions in the land surface due to thawing processes when ground ice disappears leaving thousands of shallow basins imprinted on the surface (Figs. 3.33–3.34). As the permafrost ground is frost shattered and high in ice content, waves and selective melting may open a path through which the sea can inundate the permafrost depressions (Fig. 3.33 and 3.34).

NEXT PAGE

▶Fig. 3.33 The ocean has invaded permafrost depressions forming a very irregular coastline in northern Alaska at *69°40′N* and *134°08′W*. Scene is 21 km wide. (Photo credit: ©Google Earth 2010).

▶Fig. 3.34 Closed water-filled depressions, partly flooded and opened to the sea, show similar forms to sink holes but are meltwater ponds in decaying permafrost terrain at the north coast of Alaska at about *70°N* and *130°25′W*. Width of the image is about 40 km. (Photo credit: ©Google Earth 2010).

Reference

Bird ECF (1993) The Coast of Victoria. Melbourne, Melbourne University Press.

Fig. 3.32 Water filled basins in a permafrost area of northern Alaska at *71°04′N* and *155°28′W*, 42 km wide. (Photo credit: ©Google Earth 2010).

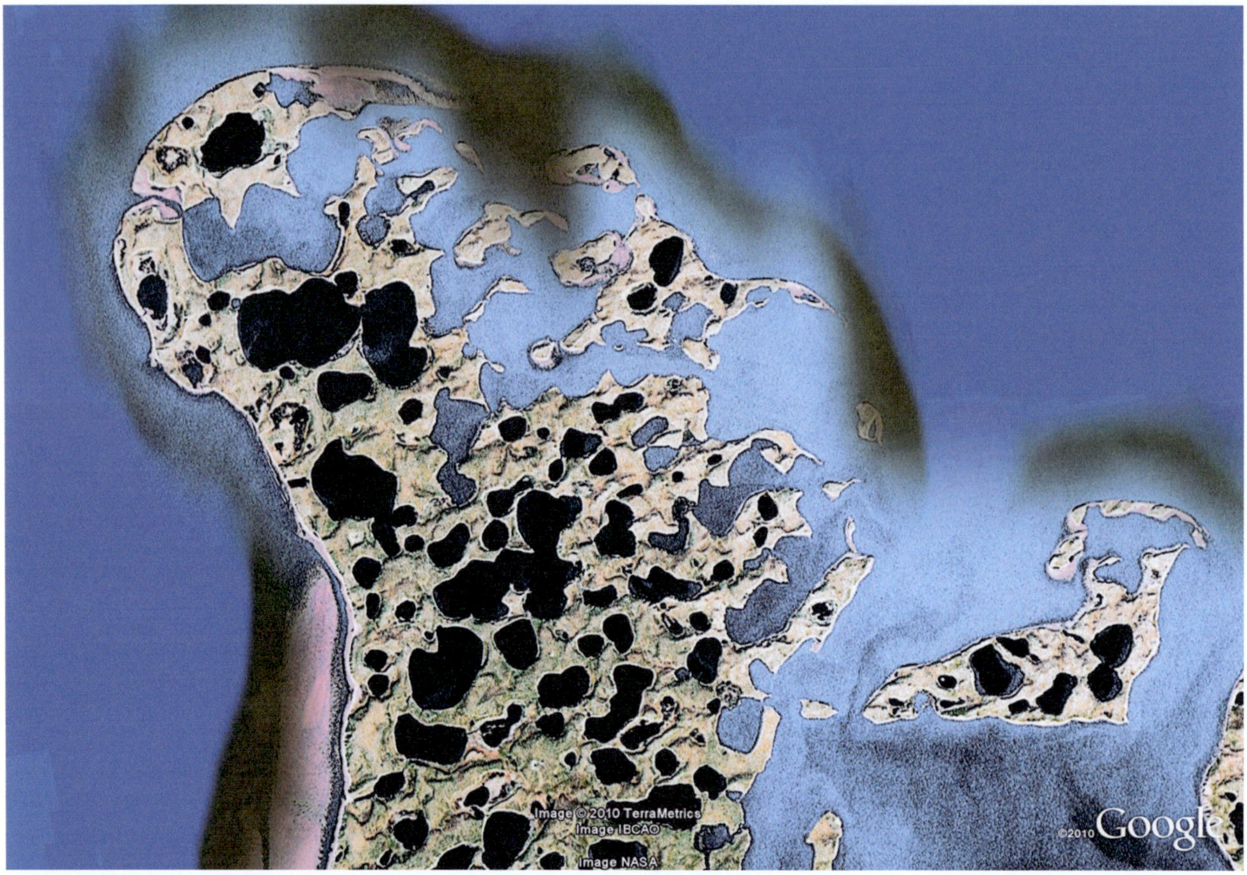
Image © 2010 TerraMetrics
Image IBCAO
Image NASA
©2010 Google

3.33

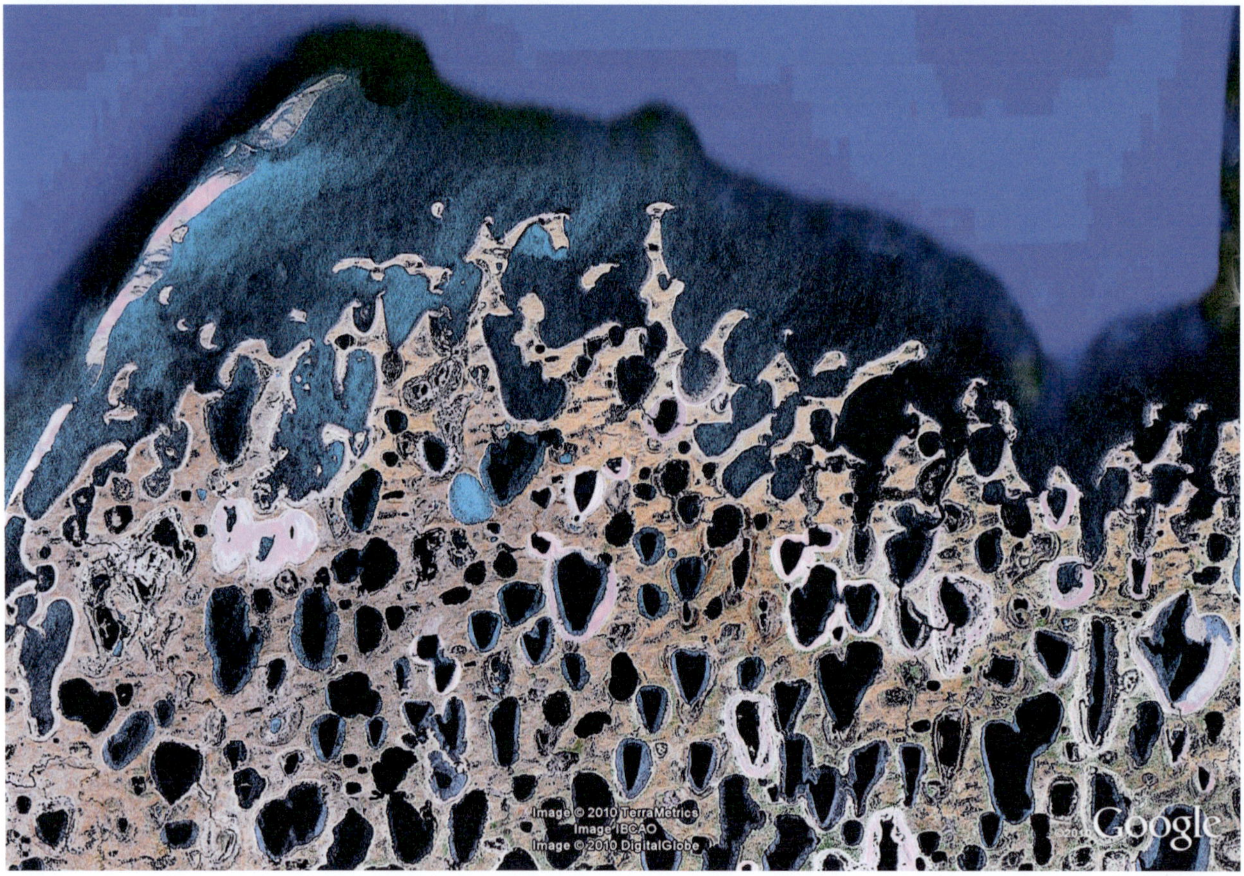
Image © 2010 TerraMetrics
Image IBCAO
Image © 2010 DigitalGlobe
Google

3.34

4

Destructive Coastlines

ABSTRACT Coastal morphology and deposits are dependent on the geologic setting and history, climate, oceanographic environment and sediment supply, coastline orientation and exposure and the rate of past and modern sea level changes. Along secondary coasts the major geological forces operating at a shoreline – the line where the water surface intersects the shore – are waves and tides. Over the last 6000–7000yrs sea level has only oscillated slightly around its present day position. Therefore, for millennia waves have rolled onshore with calm troughs during quiet weather or breaking with fearful violence during storms and transforming, eroding and reshaping the coastlines around the world. In this chapter we will outline the result of abiotic destructive processes by these hydrodynamic forces from the sea and briefly discuss coastal forms produced by salt weathering. However, living organisms can affect coastal geomorphology on a grand scale with changes even perceptible on a human time scale. Here, we see how tiny organisms in the coastal zone create and shape coastal landforms and start with an important one: Bioerosion.

4.1 Bioerosion

Geologists and biologists have described many different types of bioeroding or bioconstructive organisms (or biota) including plants and animals ranging from very small to very large: algae, bacteria, corals, foraminifera, sponges, bryozoa, barnacles, gastropods, bivalves, echinoderms, fish, mammals or worms. The discipline of biogeomorphology studies the interaction between these living organisms and geomorphological processes which take place in all climate zones and marine environments – from the rocky intertidal zone or coral reefs down to the top of deep sea knolls. Plants and animals are tuned to specific environments and will therefore thrive in certain locations. This is also the case in the coastal zone where coastal geomorphology and geomorphological processes define environmental gradients between high and low, wet and dry and sedimentation or erosion. These gradients reflect different exposure to wave impacts, nutrient levels, and abundance of organic matter or moisture. The distribution and zonation of marine boring organisms (and those of bioconstructive organisms as well as we see later) reflect the ecological demands of the organism.

Fig. 4.1 Along the microtidal coastline of southern Crete (Greece) bioeroding organisms of the intertidal zone abrade these notches (0.6 m high). (Photo credit: D. Kelletat).

Fig. 4.2a,b,c,d If sea-level is fairly constant over several millennia, bio-erosive notches may be incised many meters into limestone. a) Bonaire Island, southern Caribbean, b) Palau Rock Islands, Micronesia; c) threefold notch in western Thailand; d) mushroom rock at the Abrolhos Islands, western Australia. (Photo credit: A. Scheffers).

On rocky shores and coral reefs a typical community of grazing, burrowing or boring organisms influenced by abiotic factors such as wave energy, splash water, inundation frequency and period and depth or height in relation to present sea level effectively erodes the substrate of the rocks in certain depth-defined habitats (Figs. 4.1 to 4.17). Some organisms dwell on the surface of the underlying substrate in search for food (e.g. gastropods, sea urchins) and others live more or less protected in their domiciles within the substrate (boring sponges, bivalves, polychaetes). Their effect on bioerosion of the substrate is divided into biological corrosion, a chemical-corrosive process mostly from cyanobacteria or chlorophytes as well as polychaetes and sponges, and biological abrasion which results from mechanical activities such as rasping and boring which generates an erosion product in form of fine-grained detritus (Kelletat, 1997, 2005). Only a limited array of coastal borers can tackle all types of substrate. In hard substrates such as crystalline rocks or siliclastic sediments sea urchins are the most effective bioeroders. They can remove 1–10 mm of substrate per year (or 2 kg/m^2a) as filter feeders and in their aim to seek shelter from the pounding surf or predators (Allouc et al., 1996). You may find their borings also in other rock types such as tephra, basalt or sandstone (Figs. 4.14, 4.15).

Fig. 4.3 Beside notches, bioerosion may leave "trottoirs" as destructive features along limestone coasts (SE Cyprus). (Photo credit: D. Kelletat).

Fig. 4.4 In the supralittoral (the splash and spray zone), littorinids form rock pool with diameters to more than 2 m (Crete Island, Greece). (Photo credit: D. Kelletat).

Fig. 4.5 Along the water's edge of a rock pool the littorinids (up to 40,000/m) graze in the moist zone and not under water, which enlarges the pools laterally. (Photo credit: D. Kelletat).

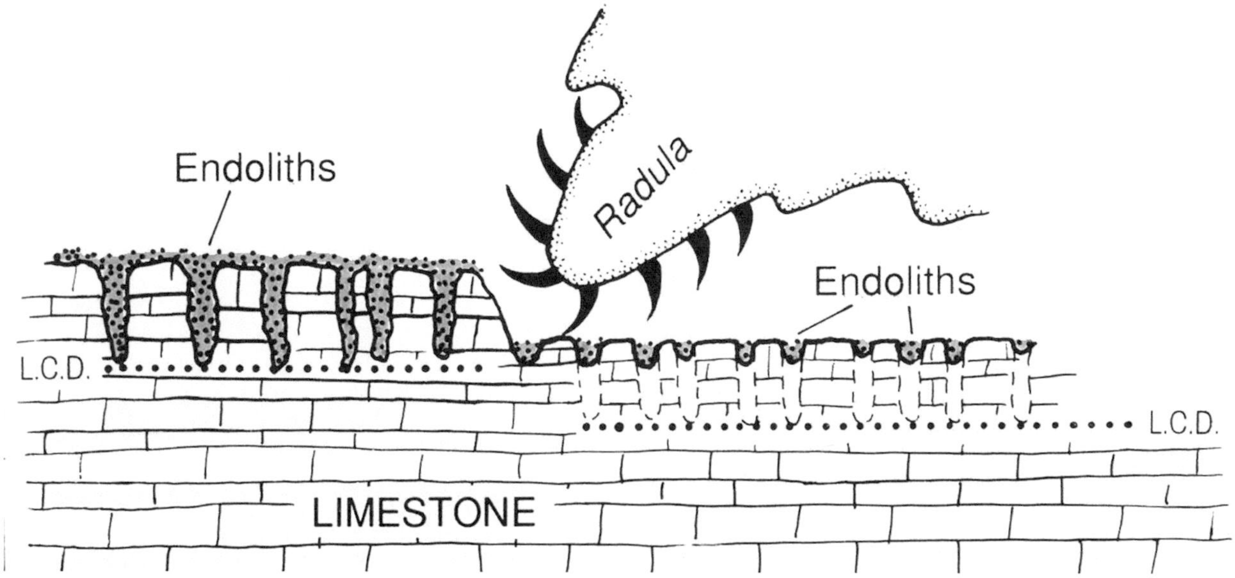

Fig. 4.6 Littorinids and the limpet use their radula (a train of teeth) for rasping away the rock and graze on endolithic algae, thus creating a new light compensation depth (L.C.D.) that allows the algae to penetrate deeper in the limestone.

Fig. 4.7 The high number of tiny littorinds result in significant bioerosion along limestone coasts. This is an example from Crete Island (Greece), where by subsidence the splash zone now reaches a higher level and has become a new grazing ground for the littorinids. (Photo credit: A. Scheffers).

Fig. 4.8 Littorinids feed on endolithic algae (the dark zone in the rock). (Photo credit: D. Kelletat).

Most significant in ecologic and sedimentologic terms is the process of bioerosion in carbonate (limestone) environments in lower latitudes such as coral reefs and along tropical limestone coasts. Bioerosion rates vary with substrate, type and agents of bioerosion between 0.5–10 mm annually. For clionid (boring) sponges (Glynn, 1997 values of 7–23 kg/m^{2}a have been noted.It is estimated that approximately 30–40% of fine sediments in reef environments is made of chips produced by marine boring sponges. Thus clionid sponges together with parrotfish may remove up to 2–3 t/ha of coral reef each year (Bromley, 1999).

Rasping bioeroders such as sea urchins, chitons and other gastropods graze on filamentous chlorophytes removing the underlying substrate more or less accidentally below the upper sub-tidal zone, thereby rapidly recycling substrate. In addition to mechanical abrasion, marine macroborers use chemicals such as calcium-complexing proteins to create their protective niche in the substrate. These organisms (among them bivalves of the family *Mytilidae* or the genus *Lithophaga*, polychaetes, sponges, barnacles) are mostly suspension feeders and their dwellings resemble the outline of the producer's body closely (Figs. 4.11–4.13). Hence, fossil

Fig. 4.9 Typical home shelters of limpets in rather hard Archaen dolomite, which are very effective bioeroders (northern Scotland). Camera cap gives scale. (Photo credit: D. Kelletat).

Fig. 4.10 The fine white stripes are scratch marks of the radula of limpets on a carpet of endolithic algae in limestone (from Cyprus). (Photo credit: D. Kelletat).

Fig. 4.11 Borings of *Lithophaga sp.* (bivalves) in sandstone (Western Australia). (Photo credit: A. Scheffers).

Fig. 4.12 The borings of bivalves in this limestone boulder in western Ireland documents its dislocation from the upper subtidal. (Photo credit: A. Scheffers).

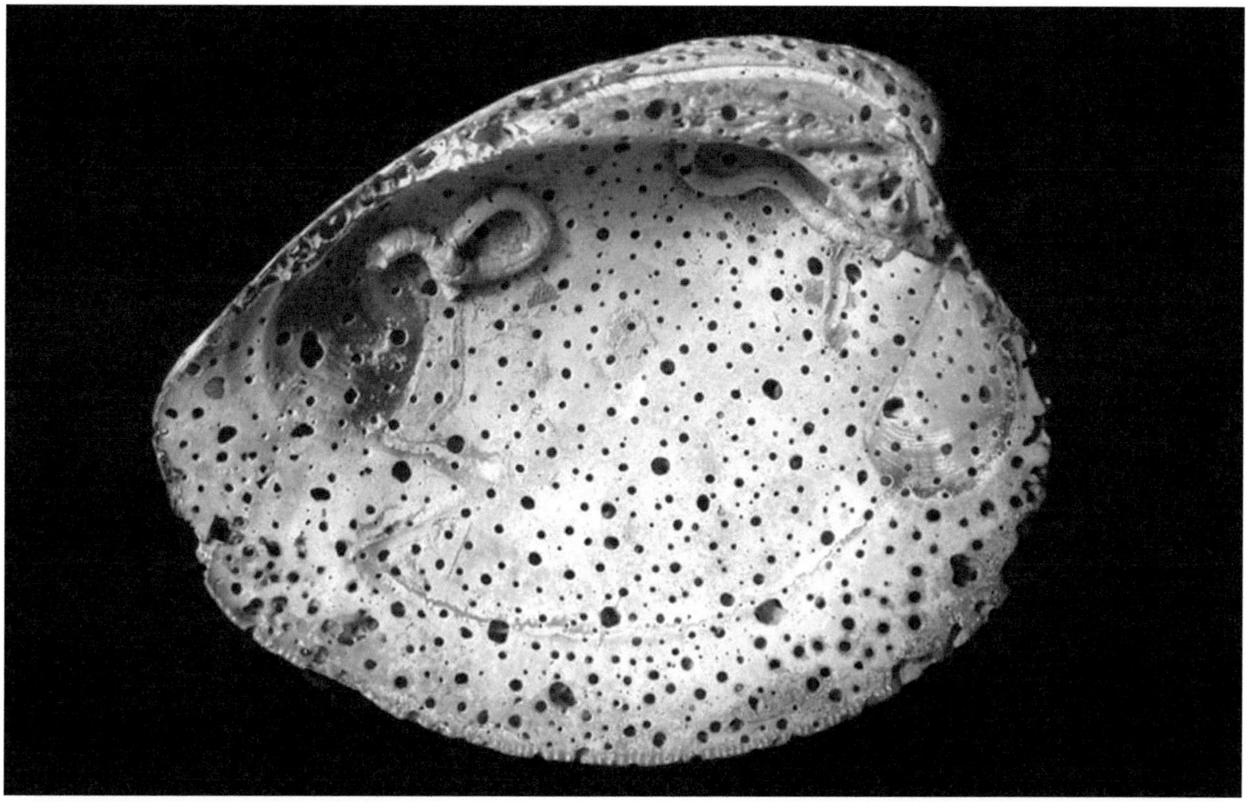

Fig. 4.13a Sponge borings (Entobia made by the genus Cliona) and encrusters on a shell of the modern hard clam, Mercenaria mercenaria, from North Carolina. This encrustation and boring happened after the death of the clam when the shell was empty. Image credit: ©Mark A. Wilson (Department of Geology, The College of Wooster).

borings are highly characteristic, sometimes down to the species level and can give palaeontologists relevant information about palaeoenvironments.

At coastlines the grazing and boring activities of bioeroders result in a diverse array of coastal forms and features in certain cases of astounding dimensions. These forms are excellent depth and/or height indicators in relation to sea level and may give coastal scientists clues about modern or past tidal ranges or sea level. The bioerosive action of grazing and boring organisms is strongest in the supratidal zone resulting in a highly sculptured and sharp-edged micro-relief, often called biokarst. Here, rock pools are typical from the upper splash to the higher surf area zone (Fig. 4.16). The process of this rock pool formation is not well documented. Rock pool development starts with a tiny fissure in the substrate creating a hospitable microenvironment for marine microborers such as chlorophytes and cyanobacteria (blue-green algae). Enough moisture from splash and spray or rain and protection from evaporation, wind or UV-radiation allow these organisms to live in high densities endolithic in the outer

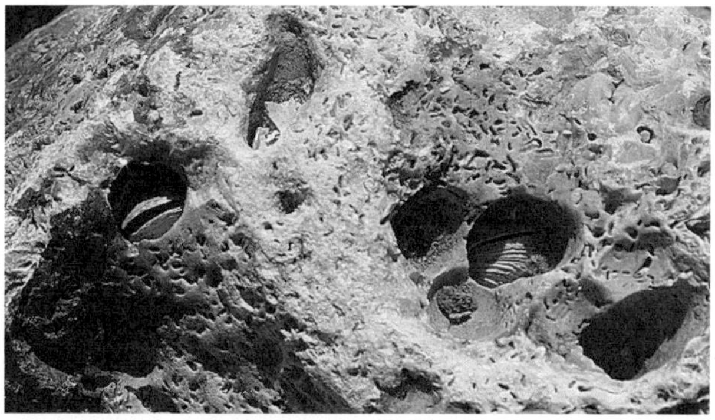

Fig. 4.13b This limestone boulder has attracted the attention of bivalves which have bored holes in it. Two holes still contain the shells of the bivalves (Lithophaga sp.) which created them. For scale, each hole is about one centimetre in diameter. (Image credit: ©Anne Burgess).

millimetres of the substrate where sunlight enables them to carry out photosynthesis. Diameters of only 5–2000 μ necessitate a Scanning Electron Microscope to see the boring perforations in the substrate which can reveal up to 800,000 per cm^2. These microboring organisms are the main food resource for gastropods (amongst them the family *Littorindae* or chitons). These sea snails use

Fig. 4.14 Borings of sea urchins in limestone of the Mediterranean, a couple of centimetres deep. (Photo credit: A. Scheffers).

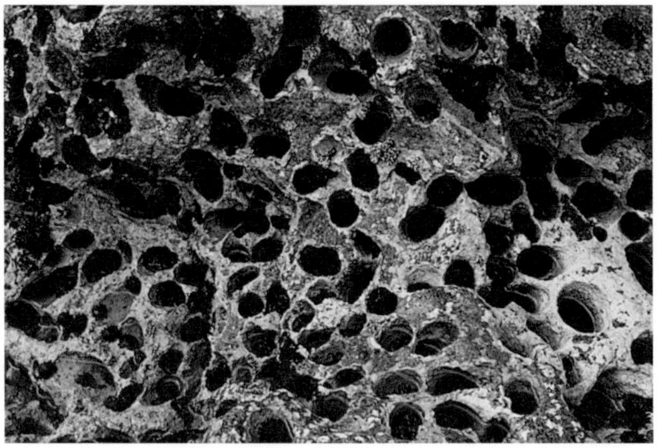

Fig. 4.15 Deep borings of sea urchins in hard basalt on Hawaii. The diameter of the bore holes is about 5–8 cm, the depth up to 25 cm. (Photo credit: A. Scheffers).

Fig. 4.16 Sharp-edged micro-topography in the spray and splash zone on limestone coasts (Bonaire, southern Caribbean). (Photo credit: A. Scheffers).

their radula (a minutely toothed, chitinous ribbon used by molluscs for feeding by scraping food off rock surfaces or other substrates) to abrade the outer rock surface to access the micro-algae as food. The abraded chips are excreted as fine-grained detritus. Successive generations of this process will produce a depression in the substrate in which the bottom is constantly covered with water. As the gastropods will not graze under water, they will concentrate their search for food and bioerosive activities along the outer rims of the depression, just a bit above the water level of

Forming and living zones from Mediterranean limestone coasts

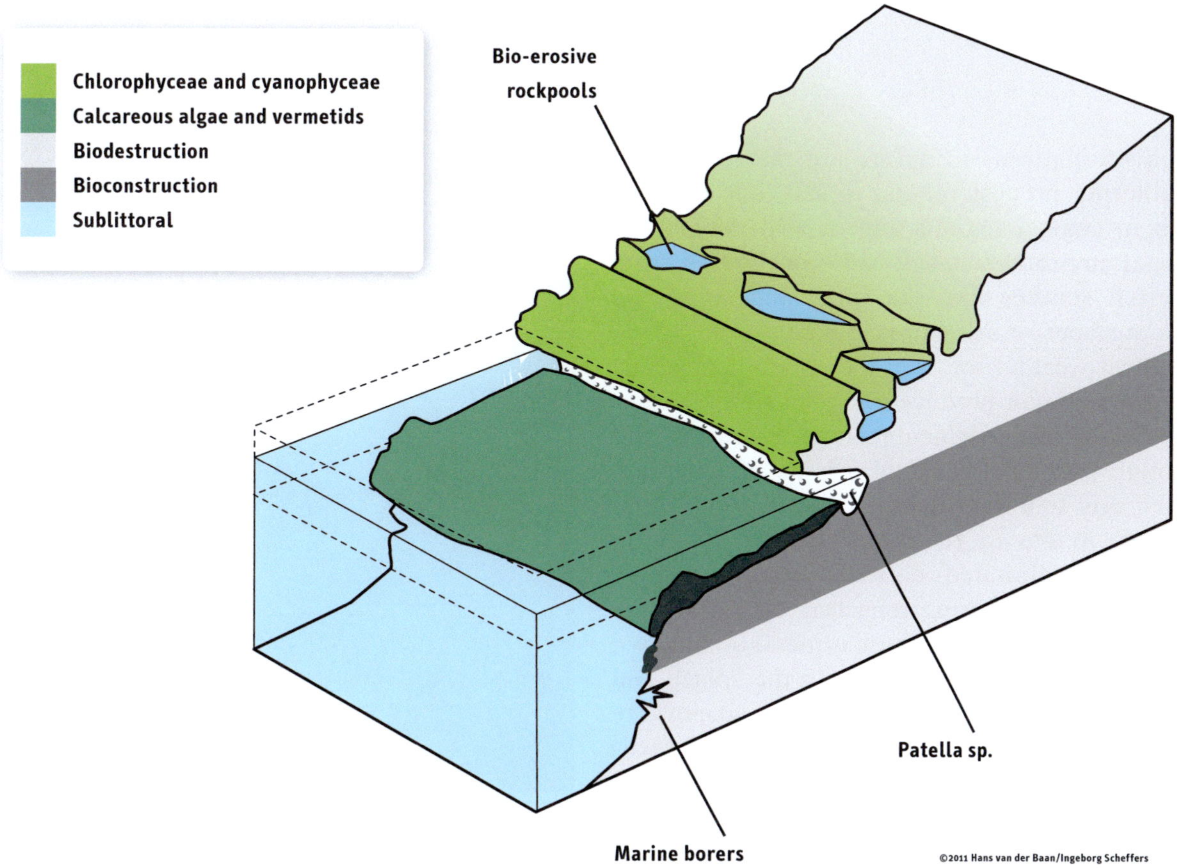

Fig. 4.17 Zones of habitats for organisms actively carrying out geomorphological forming processes, an example from Mediterranean limestone coasts. The vertical scale represents about 4–5 m, the horizontal scale may reach – depending on the slope of the rocky coast – up to 50 m (modified from Kelletat, 1999).

the initial rock pool. This will lead to a widening and lateral extension of the pool structure producing a flat pool bottom with a small notch in the outer walls. By lateral extension, neighbouring pools may coalescent, leaving one pool to dry out at the bottom and thus initializing a new rock pool generation to form along tiny little fissures at the pool bottom. Over time, a fantastic wild framework of different rock pool generations will form in all sizes and on different height levels.

Bioerosive notches are an eye-catching feature of the intertidal zone along carbonate coastlines in low latitudes. They develop as strictly horizontal back-carving incisions of a rocky shore face into steeper coastal slopes close to sea level, very often along extensive and continuous stretches of the coastline. Their lower part is located below sea level whereas the upper, larger part together with their roof is above mean sea level. Notches are excellent and precise sea-level indicators if their profile compared to the local tides is well interpreted. They appear more open and wide with a tidal range larger than 1 m and more intensive surf, and narrower with a nearly flat roof in sheltered environments with small tidal ranges. In such microtidal environments (such as on some tropical limestone islands) the notches can be incised for over 5 m into the coastal rock formation.

Organisms that are actively forming notches through abrasion and boring in these constant wet but well illuminated environments include chitons and limpets (e.g. *Patella sp.*) that graze on microboring algae. Geologists use the position and inclination of notches to determine accurately past sea level changes and/or neotectonic movements along coastlines.

4.2 Tafoni and Tessellated Pavements

Exquisite patterns in nature are the result of a weathering process related to salt that is found both in coastal regions and in arid deserts. In coastal environments salt can be delivered from external sources like sea spray and splash on rock surfaces or salt can originate internally via chemical weathering from the rocks themselves. Evaporation can produce a supersaturated solution which can permeate along fractures and through pores between the grains of the rock. When this solution finally evaporates, salt crystals precipitate in pores spaces. The resulting crystals precipitated between mineral grains exert stress and during repeating cycles of wetting and drying readily cause mineral breakdown and disintegration of the rock in the splash and spray zone of waves. Wind then can transport the resulting debris. Over time, a fascinating cavernous weathering feature develops, termed tafoni (Fig. 4.18). The word "tafoni" is thought to stem from a Corsican (French) word, *taffoni*, meaning windows, or from *tafonare* meaning to perforate. The geological term *tafoni* refers to a unique class of cavernous weathering landforms and structures. These ellipsoidal, pan- to bowl-shaped, natural rock cavities include tiny pits, softball-sized cavities, truck-sized caves, and nested and cellular honeycomb forms. Tafoni typically develop on inclined or vertical surfaces and occur in groups and may be on different rock types. Basal tafoni form at the base of outcrops and boulders where the rock comes in contact with the overlying soil (Fig. 4.19). Beside coastal environments, tafoni can be found in desert landscapes. On geologic time-scales, tafoni can cause rapid coastal landscape retreat and on human time-scales they are responsible for the destruction of stone monuments and sea walls.

Salt water also plays a role in producing some of nature's truly intriguing art work: Tessellated pavements on intertidal rock platforms (Figs. 4.20, 4.21). A tessellated pavement is a rare feature that appears in flat sedimentary rock formations on low energy coastlines and is associated with pools of standing water through a process called water

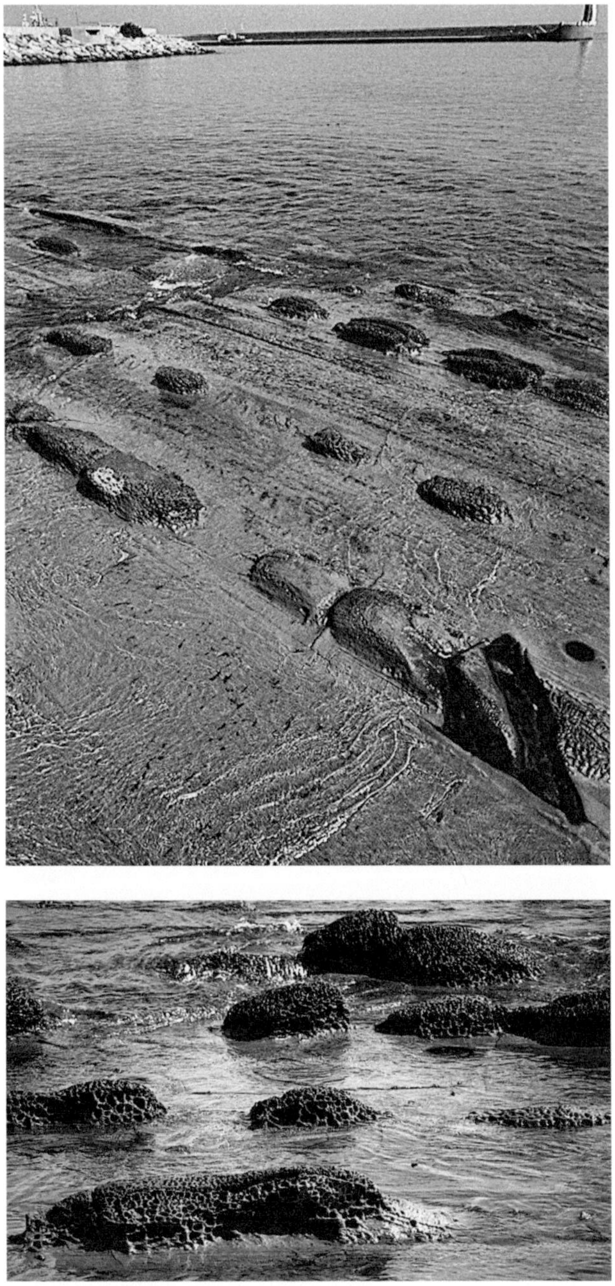

Fig. 4.18 Sandstone platforms which are exposed to alternatively wetting and drying will weather away with honeycomb forms (near Tarifa, southernmost Spain). A horizontal platform will be the result of this process. (Photo credit: D. Kelletat).

layer weathering. This type of weathering includes alternate wetting and drying, salt crystallization and many other chemical weathering processes. The pavements bear this name because the rock has fractured into regular rectangular blocks that resemble tiles, or tessellations. The cracks were formed when the rock fractured through tension and stress in the Earth's crust and subsequently were modified by wave action and erosion. The

development of such coastal features is beautifully illustrated for the Tessellated Pavement in south east Tasmania (Australia): The flat-lying siltstone was cracked or jointed in a regular pattern by stresses in the Earth's crust, possibly between 160 million years ago and 60 million years ago. This jointing, exaggerated by processes of erosion, has created the "tiled" appearance. When seawater covers the rock platform fragments of rock are carried away. Near the seaward edge of the platform, sand is the main cause of the erosion. When combined with wave action the erosional process causes "loaf" or "pan" formations. Away from the seashore the pavement dries out for longer periods at low tide and this allows greater development of salt crystals. The salt forms on the surface and erodes the pavement's surface more quickly than the joints. The surface of the pavement is lowered, while the "joints", which erode more slowly, become rims. These "pans" contrast with the "loafs', where the joints erode more quickly than the surface, because of abrasion by sand and other particles carried by water. The "loaf' formations have eroded down to a harder bed of mudstone. The joints in the lower bed are visible and sea water, sand and silt act upon the cracks. Even before the old breadloaves have been dislodged by wave action a new cycle of erosion begins on the next layer of the Tessellated Pavement. The "loaf" features are closer to the sea and so spend a longer time under water. As the drying period is shorter, salt crystallisation is less significant than that further inland. Sediment, such as sand, carried by water is the main form of abrasion. The joints tend to channel the water and the mar-

Fig. 4.19a,b Large granite boulders near Esperance, southern West Australia affected by basal tafoni weathering. (Photo credit: A. Scheffers).

gins of the blocks are eroded and the loaf tops thus appear to "rise" above the platform. At the shoreline, wave action breaks the "loaves" from the joints (Parks and Wildlife Service Tasmania, Exerpt from: Geodiversity, Tasman Peninsula, http://www.parks.tas.gov.au/file.aspx?id=7038).

Fig. 4.20 A horizontal intertidal rock platform at the "Tessellated Pavement" in SE Tasmania, Australia, as a result of water layer or salt weathering. (Photo credit: A. Scheffers).

4.3 Cliffs and Shore Platforms

Cliff coasts with their striking façades and adjacent platforms often designated as geoheritage features as they are amongst the most valued natural landscapes and they may have great symbolic value for the people. There are many poems and songs that allude to the mystery and romance of waves crashing against rocky cliffs: for example the iconic White Cliffs of Dover of Britain, facing the Strait of Dover and France, composed of soft, white chalk, have inspired movies from James Bond to Robin Hood, poetry to express English exiles' homesickness as they were the last glimpse of England while crossing the English channel to Europe's mainland, and diverse music featuring reggae or rock.

Fig. 4.21 Quarrying by wave action on the weathered rock platform of the "Tessellated Pavement" of SE Tasmania. (Photo credit: A. Scheffers).

However, geologists and geomorphologists are still puzzling over the details of the evolution and genetic origin of cliffs. The debate mainly circles around the relative roles of mechanical wave abrasion and subaerial weathering processes such as the above mentioned water layer weathering in the development of shore platforms. We here use the term shore platforms as it is most widely used by coastal scientists and has no reference to their genetic origin. Other synonymous terms include abrasion or intertidal platform, wave-cut or surf-cut terrace, schorre or simply rock bench. Shore platforms are horizontal or gently sloping surfaces cutting across geologic structures and rock formations and are backed at their landward extension by a cliff (Figs. 4.23–4.26). Cliff and shore platforms normally occur together, however some cliffs plunge directly into deeper water. These develop where the postglacial sea level rise was

Fig. 4.22 Basalt columns cut horizontally by wave action in Victoria, Australia. (Photo credit: A. Scheffers).

Fig. 4.23 Three examples of shore platforms developed in hard metamorphic rocks in the St. Lawrence River, Quebec, Canada. These were cut by floating ice and debris loaded waves. (Photo credit: D. Kelletat).

much greater than the rate of sediment accumulation at the foot of the cliff (Figs. 4.27–4.32).

The development of cliffs and shore platforms and their morphological characteristics depend on the underlying geology, the type of rock formations (rock strength) occurring along the coast, the overall wave climate and energy and other oceanographic factors such as tidal ranges. As the sea pounds rocky shores, cliff faces and shore platforms face the main force of mechanical wave erosion This includes the compression of air in joints and fissures in the rock. Air compression results from the sudden inrush of water into a joint or cavity and causes an explosive shock through the sudden release of compressed air. This process is very effective even during fair weather wave conditions since water depth determines the position of breaking waves relative to the shoreline. Larger storm waves break in deeper water further away from the shore and shoal across the platform

▸Fig. 4.25a,b Shore platforms exposed during low water along the macro-tidal coastline of western France. (Photo credit: ©Google Earth 2010).

Fig. 4.24 Inclined rock strata on a shore platform on the west coast of Ireland. (Photo credit: D. Kelletat)

Image © 2010 IGN-France
© 2010 Cnes/Spot Image
©2010 Google
4.25a

© 2010 Cnes/Spot Image
Image © 2010 IGN-France
©2010 Google
4.25b

Isle North Ronaldsay exhibits numerous cliffs and inclining abrasional rock platforms

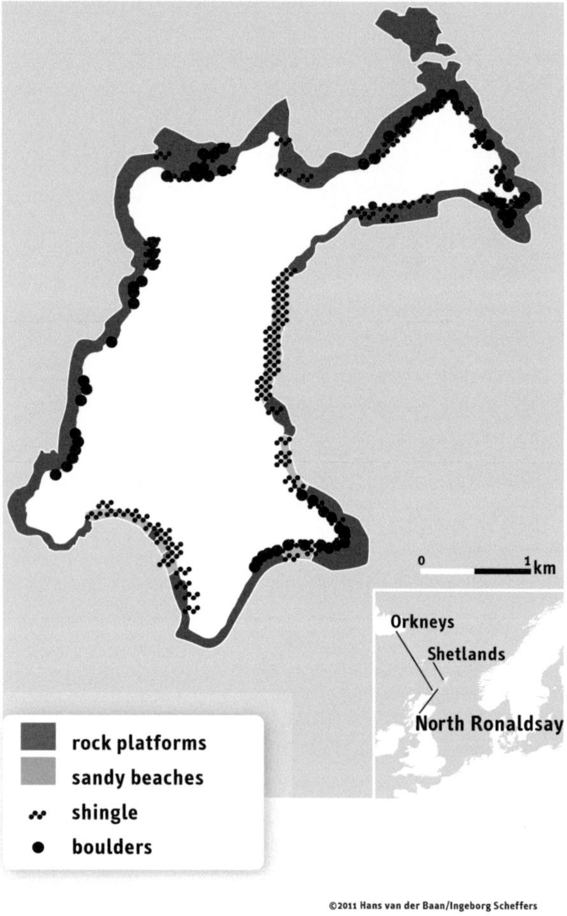

Fig. 4.26 The island of North Ronaldsay of the Orkneys (Scotland) is surrounded by cliffs and an inclined abrasional shore platform

surface rather than directly impacting against the cliffs. Associated with mechanical wave erosion is the process of abrasion, which is the scouring and wearing away under wave action or tidal currents of the underlying bedrock by the movement of sediments ranging in size from silt to boulders. If sediments generated by abrasion are not transported away by currents and tides, they may accumulate on the shore platform and further erosion will cease. In higher latitudes ice grounding by wind and waves is also an important process in platform development.

Cliffs show a vast catalogue of accompanying features, which often exist only for moments in geologic time (Figs 4.27–4.47): Small bays, narrow inlets (geos), caves, arches or stacks are usually formed by erosion along structural weaknesses in the rock formation. At the cliff foot within the reach of surf, notches with smooth contours are abraded and polished by sediment loaded waves. The deepest part of the notch often is situated a little below the mean high tide level. The occurrence and morphology of cliff caves, tunnels and arches is often controlled by weaknesses of the rock due to joints, fractures, faults or other geologic inheritances such as unconformities or changes in sedimentation patterns. It is thought that arches develop through the coalescence of caves eroded in the opposite sides of headlands. Spectacular coastal landforms are sea stacks, blocks of erosion-resistant rock isolated from the land by sea. Their existence begin as part of a headland or sea cliff, however over time relentless pounding by waves erodes the softer, weaker parts of a rock first, leaving harder, more resistant rocks as sea stacks isolated from the coast.

Historical accounts (spanning centuries in certain regions) or old photographs suggest that the average life span of sea stacks or natural bridges from their initial separation to their eventual collapse is somewhere between 100–250yrs. Overall, the erosion rate of sea cliffs is rather episodic and site-specific. Reported rates vary from virtually nothing to up to 100m/year depending on the rock strength in regard to resistance to erosion and the frequency of waves exceeding the minimum height capable of erosion (Sunamura, 1992). The persistence of some cliff coasts is well documented by the existence of ancient coastal forts from Neolithic to Iron Age times or Medieval castles in many regions of the world that have withstood wave action for centuries due to the resistant rocks on which theywere built. Cliff profiles are strongly influenced by the type of rock (harder versus softerr), structural weaknesses, stratigraphic variations or the orientation of the strata. Cliffs may also occur in soft rock formations or unconsolidated material, along coastal dunes or glacial and fluvial deposits, but they are cut back quickly and loose steepness by rain or groundwater seepage (Figs. 4.48–4.51).

▶Fig. 4.27 40 m high cliffs in limestone in SW Portugal. (Photo credit: D. Kelletat).

▶Fig. 4.28 A 5 km long cliff section at the Atlantic coast of Morocco at *28° 13′N* and *11° 45′W*. (Photo credit: ©Google Earth 2010).

4.27

Image © 2010 DigitalGlobe
Data SIO, NOAA, U.S. Navy, NGA, GEBCO
© 2010 Cnes/Spot Image
Image NASA
©2010 Google

4.28

Fig. 4.29a,b,c,d High cliffs with coarse clast cliff top deposits at (a) Moher, western Island (b,c,d) the Aran Islands, Ireland. The origin of the deposits either from storms or tsunamis is under debate. (Photo credit: A. Scheffers).

Fig. 4.30 A mega-cliff (about 300 m high) at the south coast of Gran Canaria, Spain. (Photo credit: D. Kelletat).

Fig. 4.31 These fresh break-outs of an old coral reef rock (several 100 tonnes each) on the island of Bonaire (Caribbean) are the result of 12 m waves from hurricane Ivan in 2004. (Photo credit: A. Scheffers).

Fig. 4.32 Cliff collapse along vertical joints may also occur without the influence of strong wave action, just by normal weathering, in particular where frost is present during the winter. West coast of Inishmore Island, Aran Islands, Ireland. (Photo credit: A. Scheffers).

Fig. 4.33 Quarrying of well stratified limestone rocks at cliffs in western Ireland. (Photo credit: D. Kelletat).

Fig. 4.34 Quarrying along a network of joints and fissures in coastal rocks of the Galway Bay, western Ireland. (Photo credit: D. Kelletat).

Fig. 4.35 The Aran Islands partly show undermining of slightly seaward dipping limestone strata with accelerated cliff recession forming cliff bays. Inishmore, western Ireland. (Photo credit: A. Scheffers).

Fig. 4.36 Dominant joint patterns and directions are easily eroded by strong waves like in the dolerite of southern Tasmania, Australia. (Photo credit: D. Kelletat).

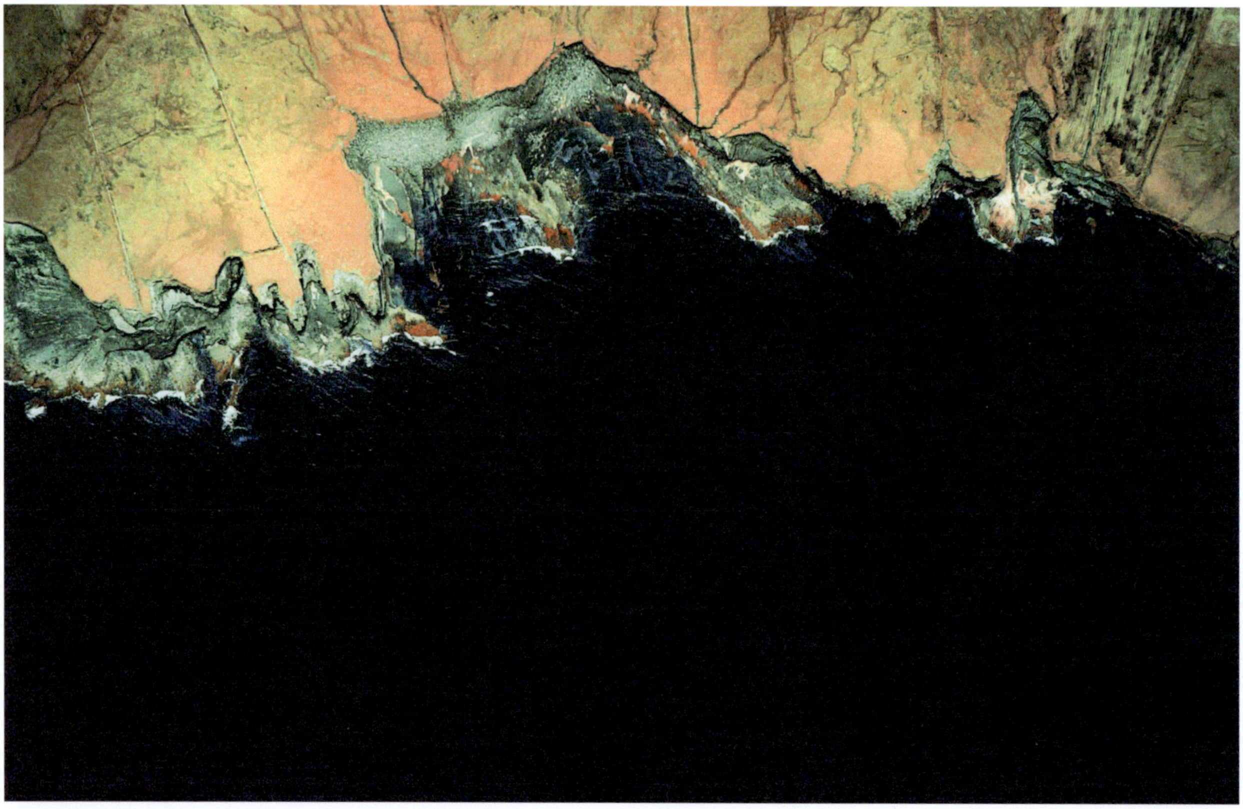

Fig. 4.37 Ongoing cliff erosion and formation of cliff bays with stacks and more sheltered places with initial pocket beaches (western Scotland). (Photo credit: D. Kelletat).

Fig. 4.38 A pocket beach dominated by cliffs where erosion is still active during intervals of high water and large waves, Shetland Islands. (Photo credit: D. Kelletat).

Fig. 4.39 Cliffs, stacks and abrasion directed by parallel joints along the "Great Ocean Road" in Victoria, Australia. Position is about *38° 39′S* and *143° 04′E*. Scene is 2.4 km wide. (Photo credit: ©Google Earth 2010).

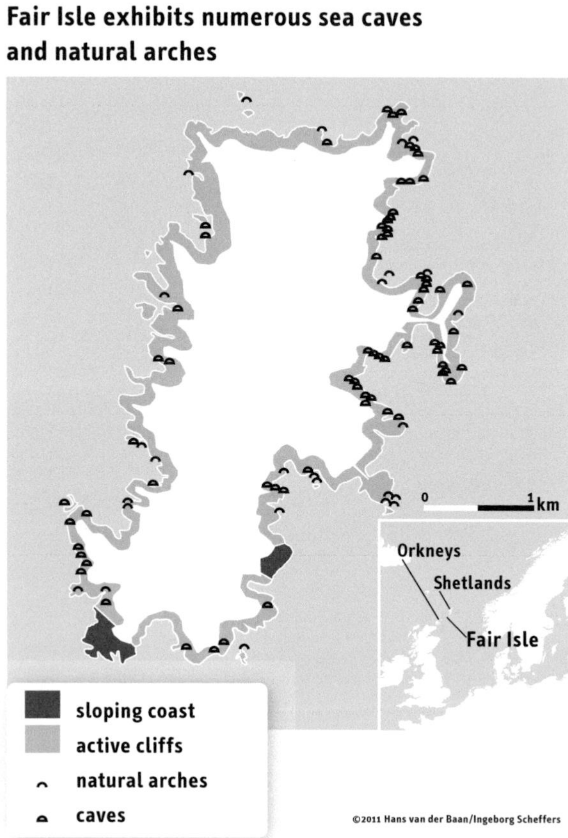

Fig. 4.40 Fair Isle between the Orkneys and Shetland Islands of Scotland exhibit numerous sea caves and natural arches. (Image credit: D. Kelletat).

Fig. 4.41 A stack about 30 m high in sandstone of Helgoland, Germany. (Photo credit: D. Kelletat).

Fig. 4.42a

Fig. 4.42a,b Stacks away from the active cliffs document a wide area of erosion/abrasion in these shales of western Ireland. (Photo credit: D. Kelletat).

Fig. 4.43 A natural arch formed by abrasional forces on granite near Albany (Western Australia). (Photo credit: A. Scheffers).

Fig. 4.44 The "Azur Window" of Gozo Island, Maltese islands, Mediterranean, adapted to the rock strata. (Photo credit: D. Kelletat).

Fig. 4.45 A natural arch in a small island in the Shetland group, northern Scotland. (Photo credit: D. Kelletat).

Fig. 4.46 The so called "London Bridge" at the Great Ocean Road in Victoria, Australia, collapsed in the year 1990. (Photo credit: D. Kelletat).

Fig. 4.47 A sea stack and natural arch in chert limestone of the white cliffs of French Normandy coast. (Photo credit: D. Kelletat).

►Fig. 4.49 Cliffs in soft rock formation cut by gullies due to heavy rain at the east coast of Brazil at *9° 53′S* and *35° 55′W.* Coastal section shown is 1.3 km long. (Photo credit: ©Google Earth 2010).

►Fig. 4.50 Cliffs in unconsolidated deposits show slides and mud flows (eastern England at *54° 34′N* and *0° 50′W*, about 3 km long section). (Photo credit: ©Google Earth 2010).

Fig. 4.48 Two examples of unconsolidated younger sediments with cliffs and slumps along the coast of Algarve, southern Portugal. (Photo credit: D. Kelletat).

Image © 2010 DigitalGlobe
©2010 Google

4.49

Image © 2010 TerraMetrics
© 2010 Infoterra Ltd & Bluesky
Image © 2010 The GeoInformation Group
©2010 Google

4.50

Fig. 4.51 Cliffs may also exist for a short time as quasi stable features in sand as in these mid-Holocene dunes with soil development and vegetation cover of Sylt Island, German North Sea coast. Many houses have been destroyed by rapid cliff retreat in these dunes during the last 200 years. (Photo credit: D. Kelletat).

References

Allouc et al. (1996) La bioérosion des substrats magmatiques en milieu littoral: l'exemple de la presqu'ile du Cap Vert (Sénégal occidental) Géobios 29:485–502.

Bromley RG (1999) Bioerosion: a collection of workshop papers. Bulletin of the Geological Society of Denmark Marine Geology, 44: 129–180.

Glynn PW (1997) Bioerosion and coral reef growth; a dynamic balance; in Birkeland (ed.) Life and Death of Coral Reefs; Chapman & Hall, International Thomson Publ. New York, USA.

Kelletat DH (1995) Atlas of Coastal Geomorphology and Zonality. Journal of Coastal Research, Special Issue No 13.

Kelletat D (1997) Mediterranean coastal biogeomorphology: processes, forms and sea-level indicators. In: Briand F, Maldonado A (ed) Transformations and evolution of the Mediterranean coastline. CIESM Science Series No 3, Bulletin de l´Institut Océanographique, Monaco Numéro Special 18:209–226.

Kelletat D (1999) Physische Geographie der Meere und Küsten. Teubner Studienbücher Geographie, 2 ed., Stuttgart

Kelletat D (2005) Schwartz ML (Ed.) Encyclopedia of Coastal Science. Springer, Dordrecht.

Short AD, Woodroffe CD (2009) The Coast of Australia. Cambridge University Press, Cambridge.

Stephenson WJ (2000) Shore platforms remain a neglected coastal feature. Progress in Physical Geography, 24: 311–327.

Sunamura T (1992) Geomorphology of Rocky Coasts. Chichester, John Wiley.

5

Sedimentary Coasts

ABSTRACT The most popular of coastlines are beaches. Beaches are primary landforms bounding approximately 30% of the world's coastlines and consist of sand or gravel (or a mixture of both). They have enormous recreational value but importantly act as buffers for wave energy delivered to the shore and shelter areas behind the beach from wave attack or flooding, especially during storms. Along many coastlines, beaches and the associated back-beach environments (lagoons, marshes, dune belts) have been intensively developed and are today among the most densely populated regions of the world and therefore particularly vulnerable to the impacts of marine natural hazards. A beach is a three-dimensional sediment body along a shoreline that extends from the upper limits of wave run-up to the outer limits of wave action in the nearshore zone. In 124 images we demonstrate the diversity of beaches and associated often ephemeral features both on land and in the littoral zone: Dunes, spits, barriers, tombolos, beachrock, beach ridge systems and coastal landforms at the interface between rivers and the sea: different delta types of the world.

5.1 Introduction – The beach and its features

There is no standard definition of what constitutes "the beach" because it depends largely on one's perspective or the scientific question: Recreational beach users will mostly consider the extent of a beach no farther away than the shoreline and thus see the beach as an entirely emergent coastal feature which may change with different water levels. Coastal scientists often refer to the beach as "the zone containing unconsolidated material that extends from the limit of mean low-tide level on its seaward side to the limit of influence by storm waves on its landward extent" (Davidson-Arnott, 2010; Short, 2006; Schwartz, ed., 2005). Many beaches are more or less straight stretches of sand that extend from 1 km to more than 100 km in length; others are simply smaller crescents of sand between rocky headlands (or pocket beaches). A beach may appear to be stable, but constantly gains and looses sediments on all sides: Waves and littoral drift move sand down the beach where during the journey and at the end of the beach sand is removed and deposited in deeper water. Most sediments reach and replenish the shoreline via rivers that are carrying their sediment load to the sea. Along cliffs and in the backshore, sand or pebbles are freed by erosion and contribute to the sand budget of a beach. Also wind that blows over the beach is transporting sand – sometimes offshore into the ocean and sometimes onshore onto land. Altogether these processes maintain a fragile balance between the inputs and outputs caused by erosion and sedimentation or the sand budget of a beach. If the beach is in equilibrium, it will more or less keep the same form and shape. If the inputs and outputs are not balanced, the beach grows or shrinks. Temporarily (over days, weeks or month) this is a natural process: During storms, powerful waves erode the exposed part of a beach and

©2010 MapLink/Tele Atlas
©2010 Europa Technologies
©2010 Google

5.1

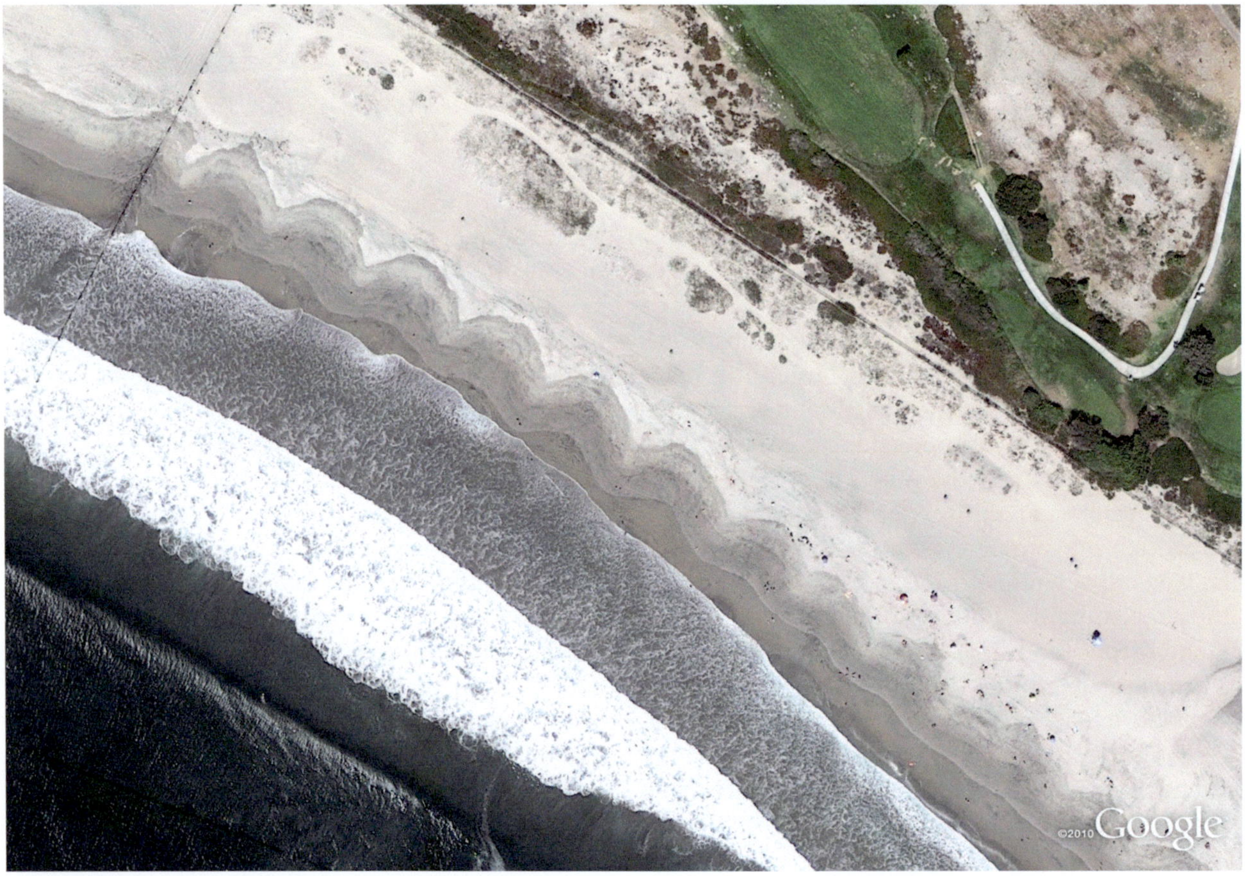
©2010 Google

5.2

◀ Fig. 5.1 Rhythmic features along beaches are called beach cusps and are the result of waves and/or currents approaching with an angle to the coastline. (Image credit: ©Google Earth 2010).

◀ Fig. 5.2 A well developed reflective beach containing high tide beach cusps at Hammer Head, south Western Australia (Photo credit: AD Short, OzCoasts; http://www.ozcoasts.org.au/).

makes it narrower. In calm, mild weather, lower waves will move sand onshore and rebuild a wider beach. Because storminess tends to be seasonal, most beaches also change character seasonally.

Beaches can be divided based on morphology in two zones: The backshore zone is the higher, more landward part of the beach usually not affected by the run-up of waves except during storm wave conditions. The landward extent is usually characterized by a change in morphology (to a beach cliff or dunes) and/or vegetation. Beach cliffs or scarps are topographic expressions of erosion during storm events causing the beachface to migrate landward by cutting into the backshore zone. They can vary in height from a few centimetres to several meters depending on the degree of wave action and the type of material. The foreshore zone is the more seaward part of the beach extending to the

Fig. 5.3 Beach cusps along a shoreline in California. Width of image about 400 m. (Image credit: ©Google Earth 2010).

Fig. 5.4 Dark layers of heavy minerals (magnetite, limonite) in stratified beach sands of Mombasa, Kenya, eastern Africa. (Photo credit: D. Kelletat).

Fig. 5.5 Beach stratification at the Mombasa coastline of Kenya, eastern Africa. (Photo credit: D. Kelletat).

limit of mean low water including the surf zone, the tidal flat and right at the shore the swash zone, a slope dominated by the swash and backwash of waves. The lower part of the foreshore is submerged during high tides. A distinguishing beach feature of the upper part of a beach is a broad, depositional horizontal to gently landward sloping area called a beach berm which is formed by onshore accumulation of sand from both average waves or in a higher position storm waves. The berm crest divides the foreshore from the backshore zone. The zone seaward from the low water shoreline is termed the nearshore zone and extends out to the breaker zone, thus this zone is constantly submerged.

In cases, the berm crest and the foreshore can feature rhythmic, crescentic patterns, the beach cusps (Figs. 5.1–5.3). Their form can vary greatly in dimensions and height. All cusps are characterized by seaward facing ridges or horns separated by bays. They can form by different mechanisms of waves and currents in the nearshore zone (Komar, 1998).

Beaches are mostly composed of sand-size sediments from mineral, rock or shell fragments and usually the beach will be made of materials that are locally available for waves to rework. In temperate climates beaches typically consist of quartz and feldspar grains derived from the weathering of terrestrial rocks but may contain a larger fraction of shells (either broken or intact) and also foraminifera. Along tropical beaches, the composition of the beach material can derive entirely from calcium carbonates from marine organisms

Fig. 5.6 A Well rounded pebbles on a modern beach in Ireland. (Photo credit: A. Scheffers).

Fig. 5.6b Rounded quartzite boulders from an ancient beach deposit in SW Ireland. (Photo credit: A. Scheffers).

Fig. 5.7 Extended pebble and cobble beaches, mostly with ridge forms in a formerly glaciated area, west coast of Ireland. (Photo credit: A. Scheffers).

Fig. 5.8 Beach sediment of shale fragments weathered in a frost environment along the coastline of northernmost Norway. (Photo credit: D. Kelletat).

Fig. 5.9 A boulder barricade in a fjord of northern Norway, formed by the push of drifting sea ice, is a rare form of a unsorted beach sediment from glacial drift. (Photo credit: D. Kelletat).

(shell, coral, foraminifera and others). Many patterns on a beach are made by sorting of finer and coarser particles, swash limits reveal coloured patterns of light quartz sand and darker heavy minerals (Figs. 5.4, 5.5). Coarser beach material includes gravel, cobbles, shingle or even boulders of different rock types (Figs. 5.6–5.9) or more tropical climates organic limestone fragments such as corals or shells. Wave abrasion rounds off the rough edges and usually beach gravel is well rounded, exemptions are regions where frost shattering provides the beach with more angular material frequently.

The transformation of sandy (or gravelly) beaches into a consolidated deposit result from lithification by calcium carbonates of sediment in the intertidal and spray zone of mainly tropical and subtropical beaches (Figs. 5.10, 5.11; Scoffin and Stoddart, 1987). However, beachrock can also form at higher latitudes and has been described from the coasts of Norway, Denmark, Poland, New

Fig. 5.10 Beachrock along the coast of southern Cyprus. The deposit has thickness of more than 3.5 m. (Photo credit: A. Scheffers).

Fig. 5.11 Drowned beachrock from the Peloponnese, Greece. (Photo credit: D. Kelletat).

Zealand or Japan. The process of beach lithification and beach rock formation is highly controversial with different mechanisms of cementation responsible at different locations: Some authors argue that cementation of the beach sediment takes place subsurface in the area of water table excursion between the neap low and high tide levels. Geomorphologists emphasize that beachrock rather is the product of cementation above the surface and occurs in the supratidal spray belt, which than may explain its different height level in regions with larger waves or greater exposure.

Beside waves, other important agents in forming and shaping a beach are tides, currents, wind and in higher latitudes coastal ice. Beaches and beach features are dynamic coastal landforms and may change from day to day, week to week, season to season, and year to year.

5.2 Foreshore Features and Tidal Flats

Some small scale features like ripples, small sandbars and troughs are inherently ephemeral and will only persist until new waves, tides, currents or wind conditions replace them with new features. Larger scale features like mega-ripples or a more complex ridge and runnel topography will take relatively longer to change over time. We start our excursion with these ever changing features in the foreshore zone. Where the shore is low-lying, sand is abundant and the tidal currents are strong extensive tidal flats can form which are exposed at low tide. Over 70% of tidal flats occur in wave-sheltered regions such as bays, estuaries or lagoons with the remaining class occur along open coasts characterised by low wave climates. Tidal currents transport mud in suspension and sand by traction. Tidal flats may be made of mud, sand, rock gravel and/or shell pavements (5.12–5.14). Often there is a zonation of sediment types with regard to grain size across the entire tidal flat, but in many instances one sediment type (such as mud or sand) is dominant in a certain height level of the tidal flat. Thus, coastal researchers may classify tidal flats according to grain size in muddy high tidal flats or sandy low tidal flats.

The morphology of sand tidal flats is largely shaped by waves and tidal currents with ridges (sandbanks, sandbars) and troughs (or runnels) more or less parallel to the shoreline as the most common features (5.19–5.22). They can be observed as one single ridge and runnel set or may occur as multiple set of systems forming a corrugated topography across the tidal flat. Often smaller scale rhythmic bedform patterns are referred to as ripples or in larger dimensions as mega-ripples (5.15–5.18; 5.24 and 5.25). Along most coasts the tidal range is no more than 2 m, however the horizontal distance over which the tidal zone extends depends on the tidal range and the slope of the sea bottom in the coastal zone. In bays, estuaries, straits and other narrow places along the coast, tidal ranges may be amplified and may reach 15 m or more, as in the Fundy Bay, eastern Canada. Tidal currents in those areas are often very rapid and can approach up to 25 km/h leaving behind forms from ripples to mega-ripples, sandbars and ridge/runnel features, tidal deltas, creeks and channels (5.26–5.29). Some of the best known tidal flats occur along the coastlines of the North Sea (the German and Dutch Wadden Sea, the Wash in south-eastern England); the Bay of Mont St. Michel in France; the Gulf of California; the Bay of Fundy in Nova Scotia or along King Sound in north-western Australia. Here, the tidal flats are dominated by siliclastic sediments such as quartz sand and quartz silt. Carbonate dominated tidal flats occur in mid- to low latitude warmer climate zones – famous examples are Andros Island of the Bahamas Islands and Shark Bay in Western Australia, UNESCO listed world heritage regions.

Muddy tidal flats are of a dark colour due to high organic content. The sedimentation of these fine particles occurs by settling of suspension load during low current velocity and tidal slack water (now to nil current velocities) in the short time between the incoming and outgoing tidal water flow.

Mud tidal flats are home to a rich diversity of different vegetation types (mangroves, salt marsh, algal mats) and to myriads of organisms (molluscs, crustacea, polychaetes, resident and nektonic fishes, birds) which have to cope with environmental conditions at the triple junction between land, sea and atmosphere. Benthic or sessile biota of tidal flats has to be adapted to saltwater conditions during tidal flooding, but may be exposed to freshwater (rain) during low tide. Surface temperatures may reach more than +50°C in summer and in some regions lower than -30°C in winter. With the ebb and flow of tides conditions can change within hours and sometimes within minutes. Therefore, many organisms of the benthos are burrowing types and live within the sediment as a strategy to encounter an environment with continuing similar conditions, thereby constantly bioturbating the sediment. The biogenic sediment structures of these organisms often are diagnostic for a specific tidal level.

Over time, the vegetation of the higher tidal flats (e.g. the halophyte Salicornia sp.) fixes and

Fig. 5.12 Tidal flats exposed during low water with a patch of *Salicornia sp.* in the semi-arid environment of eastern Patagonia (Argentina). (Photo credit: D. Kelletat).

Fig. 5.13 In tropical regions, sandy tidal flats may be inhibited by numerous crabs (e.g. *Dotilla sp.*) who distribute small sand pills radial to all sides around small vertical burrows in the sand. The higher the tide range, the deeper the crabs have to burrow to reach the water level in the sediment to allow breathing with their gills. (Photo credit: A. Scheffers).

Fig. 5.14 Temperate sandy tidal flats (here: north coast of Germany) are often decorated with billions of tiny sand mounds – excrements of the sandworm *Arenicola sp.*, a species diagnostic for low-tidal sand flats (Photo credit: D. Kelletat).

Fig. 5.15 Ripple fields (in the foreground) and larger sand bars (along the horizon) on a sandy tidal flat of the German North Sea coast. (Photo credit: D. Kelletat).

Fig. 5.16 Mega ripples on a sandy tidal flat of southern New Zealand. (Photo credit: D. Kelletat).

Fig. 5.17 Another form of mega-ripples, here very regular made by tidal currents along the south island of New Zealand. (Photo credit: D. Kelletat).

Fig. 5.18 Sand bars like mega-ripples migrating towards the beach from the shallow foreshore. The Netherlands at *51° 05′N* and *2° 31′E*, width is about 7 km. (Image credit: ©Google Earth 2010).

accumulates sediments that are deposited during high tide events or storms. By this process, parts of the tidal flat or the landward margin may grow above the normal high water level, and more plants settle and stabilize the ground. If sea level is stable, after hundreds of years this new salt marsh above the mean high water mark can grow above the highest storm levels and remains a terrestrial environment with new vegetation and soil development (Fig. 5.23).

NEXT PAGE

▶Fig. 5.19 A sandy tidal flat near Darwin in northern Australia shows different salt concentrations (white colours) on the surface due to evaporation. The dark fingering pattern is the rising tide water along small creeks. Oblique aerial photograph (Photo credit: A. Scheffers).

▶Fig. 5.20 Sand bars of different forms are typical for sandy tidal flats with strong currents as in the nearshore of the northern Netherlands′ coastline, here 25 km wide at *53° 30′N* and *6° 33′E*. (Image credit: ©Google Earth 2010).

▶Fig. 5.21 Tidal flats with a system of tidal creeks behind barrier islands of The Netherlands at *53° 12′N* and *5° 03′E*. The barrier island is nearly 20 km long. (Image credit: ©Google Earth 2010).

▶Fig. 5.22 Tidal mud flats in a tropical environment with numerous tidal creeks (western Mexico at *24° 47′N* and *108°W*, around 9 km wide). (Image credit: ©Google Earth 2010).

5.19

Image © 2010 Aerodata International Surveys
Image © 2010 TerraMetrics
Google

5.20

Image © 2010 Aerodata International Surveys
Data SIO, NOAA, U.S. Navy, NGA, GEBCO
©2010 Google

5.21

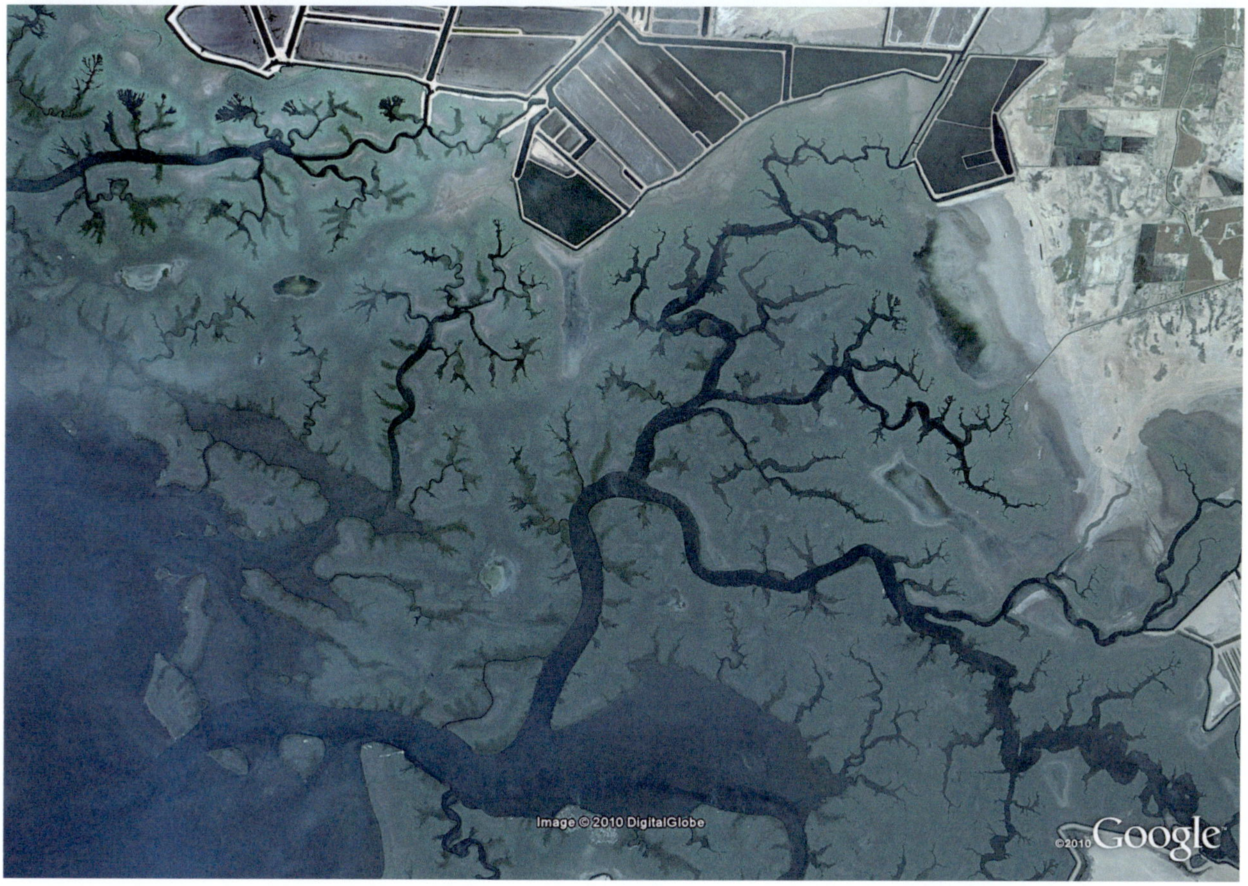
Image © 2010 DigitalGlobe
©2010 Google

5.22

Fig. 5.23 A cross-cut in marshlands show the changing pattern of light coloured sands from storm floods and dark bands of soil development. North coast of Germany (Photo credit: A. Scheffers).

►Fig. 5.25a Ripple pattern in the sandy foreshore of eastern Canada at *42°N* and *70°W*. The mega-ripple section is about 1.2 km wide, the wavelength of single ripples about 50m. (Image credit: ©Google Earth 2010).

►Fig. 5.25b Ripple pattern in the lower intra-tidal zone. Western France at *48° 40′N* and *3° 35′W*. Image shows a section 4 km wide. (Image credit: ©Google Earth 2010).

NEXT PAGES

►Fig. 5.26 Tidal currents move sands and constantly transform these deposits (eastern Canada at *42° 39′N* and *70° 44′W*, 5 km wide). (Image credit: ©Google Earth 2010).

►Fig. 5.27 A tidal delta of the US east coast at *39° 30′N* and *74° 20′W*. Scene is 9 km wide. (Image credit: ©Google Earth 2010).

►Fig. 5.28 Tidal channels between islands in a Bahamian island chain at *23° 50′N* and *76° 14′* with the deposition of the fine sediments as ebb-tidal deltas and flood-tidal deltas at both sides of the island chain. (Image credit: ©Google Earth 2010).

►Fig. 5.29 Tidal delta in a large lagoon of the north coast of the Mexican peninsula of Yucatan at *18° 45′N* and *91° 27′W*. Scene in the image is about 22 km wide. (Image credit: ©Google Earth 2010).

Fig. 5.24 Foreshore sand patterns from tidal currents in Shark Bay, Western Australia, at 26°S and 114°E in a 10 km wide section. (Image credit: ©Google Earth 2010).

Image © 2010 DigitalGlobe
Image © 2010 TerraMetrics
Data SIO, NOAA, U.S. Navy, NGA, GEBCO
©2010 Google

5.25a

Image © 2010 IGN-France
© 2010 Cnes/Spot Image
©2010 Google

5.25b

©2010 Google

5.26

Data SIO, NOAA, U.S. Navy, NGA, GEBCO
Image © 2010 TerraMetrics
©2010 Google

5.27

Image U.S. Geological Survey
Image © 2010 DigitalGlobe
©2010 Google

5.28

Image © 2010 GeoEye
Image © 2010 TerraMetrics
Data SIO, NOAA, U.S. Navy, NGA, GEBCO
Image © 2010 DigitalGlobe
©2010 Google

5.29

Fig. 5.30 A 100 m long and 3 m high coral rubble spit was formed by waves and longshore drift during hurricane Lenny within several hours in 1999 on the leeward coast of Bonaire, southern Caribbean. Drift direction was into the distance. (Photo credit: A. Scheffers).

Fig. 5.31 A complex spit at the north coast of Washington State, USA. Evidently the initial long spit has led to wave refraction and the construction of subsequent spit forms in the wave shadow of the outer one. Oblique aerial photograph, (Photo credit: D. Kelletat).

►Fig. 5.32 Spits grow from both sides of this channel in the Gulf of Mexico at *29° 11′N* and *112° 15′W*. Width of scene is 10 km. Drift and sediment transport evidently came from the north and the south alternately. (Image credit: ©Google Earth 2010).

►Fig. 5.33 A sharp contoured spit, grown from both sides at the northern tip of Queen Charlotte island, British Columbia, Canada, at *54° 07′N* and *131° 41′W*. Width of scene is 19 km. (Image credit: ©Google Earth 2010).

5.3 Spits and Tombolos

Longshore currents can transport considerable amounts of sediments along a coast and carry sand to the downcurrent end of the beach. When the coastline curves back away from the drift direction and the current encounters deeper water and its velocity decreases, the transported sediment builds up, first as a submerged bar then growing above sea level and extending the beach by a narrow addition called a spit (Figs. 5.30–5.41; as well as Long et al., 2008). These elongated sandy or gravelly ridges projecting seaward may either merely end in open water (often with curving ends) or, in cases, close off a bay and form a barrier with a lagoon behind. Spits may also congregate to form complex forelands. The variety of

Image © 2010 DigitalGlobe
Data SIO, NOAA, U.S. Navy, NGA, GEBCO
Image © 2010 TerraMetrics
©2010 Google

5.32

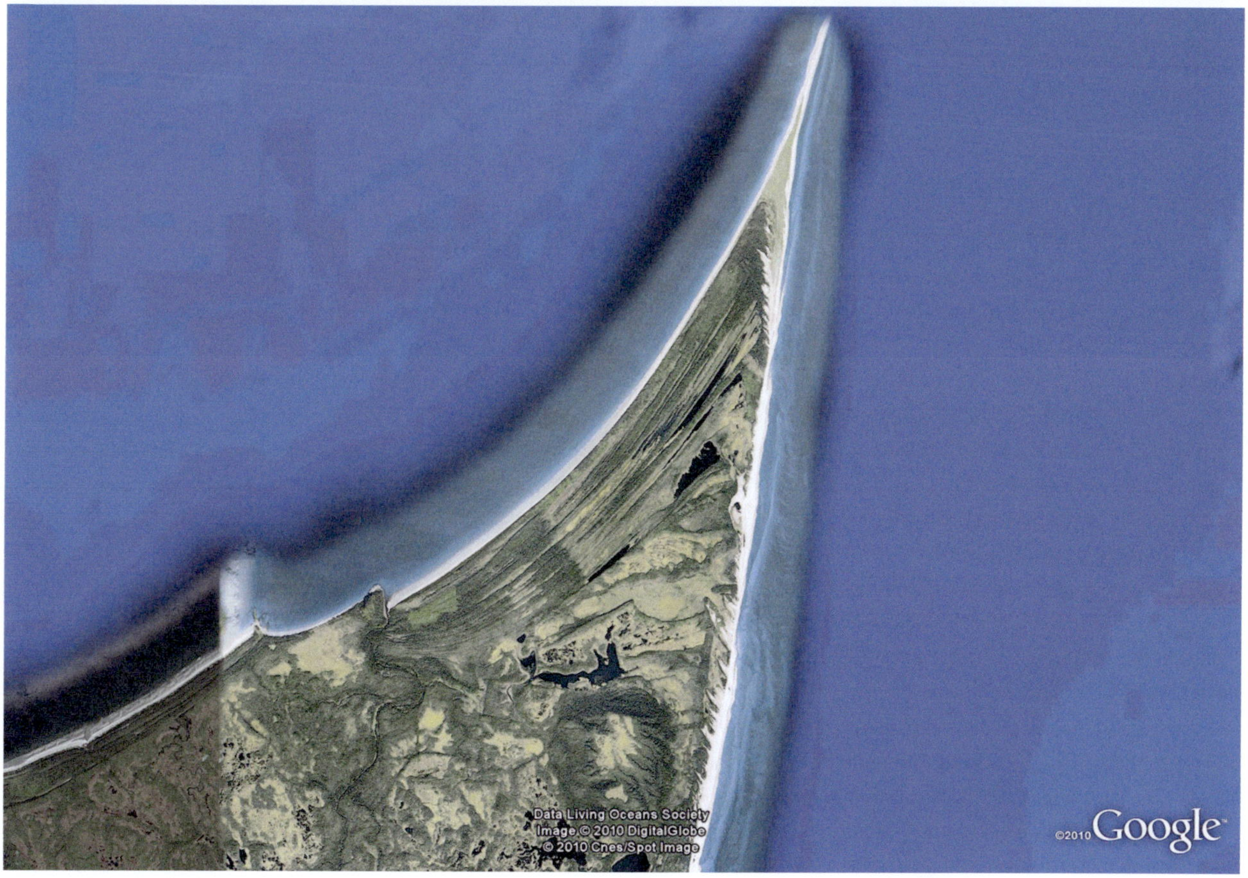
Data Living Oceans Society
Image © 2010 DigitalGlobe
© 2010 Cnes/Spot Image
©2010 Google

5.33

forms of these conspicuous coastal landforms is enormous; fine examples in large dimension occur in the Sea of Azov, or along the east coast of the USA and Canada.

A tombolo (from features in NW Italy, compare Figs. 5.42–5.46) is formed when one or more sandbars or spits connect an island to the mainland or establish a sandy bridge-like connection between two or more islands. They are formed by wave refraction around the island to the opposite side as they approached. This is causing a convergence of longshore drift on the opposite side of the island. The beach sediments that are moving by littoral drift on the lee side of the island will drop out and accumulate here. In other words, the waves sweep sediment together from both sides.

Fig. 5.34 Many steps in the growth pattern of this curved spit at the east African coast, 19°S and 35° 50′E. Width of scene is 12 km. (Image credit: ©Google Earth 2010).

NEXT PAGE

►Fig. 5.37 The tip of this spit is in danger of being cut off from the remainder (Black Sea at *45° 46′N* and *33° 11′E*, width of image 16 km). The spit has grown from two directions but in recent times the northward sediment transport along the western side dominates. (Image credit: ©Google Earth 2010).

►Fig. 5.38 A long system of curved spits in the German Baltic Sea, growing from the west. *54° 23′N* and *11°E*. Image is 4 km wide. (Image credit: ©Google Earth 2010).

►Fig. 5.39 A 14 km long curved spit system at the north coast of Hokkaido Island, Japan. (Image credit: ©Google Earth 2010).

►Fig. 5.40 This Foreland at the German Baltic Sea coast, called "Darss", has been formed during the last 5500 years at *54° 26′N* and *12° 32′E*. Width of image is 13 km. Main littoral drift is occurring from the west and southwest. (Image credit: ©Google Earth 2010).

►Fig. 5.35 In the area of this 24 km long spit longshore drift must have changed in a complex pattern. Sea of Azov northern part of Black Sea, at *46° 34′N* and *36° 22′E*. It is possible that ongoing erosion may cut the narrow central part and leave an island with a complex beach ridge system and lagoons. (Image credit: ©Google Earth 2010).

►Fig. 5.36 Curved spit 25 km long with enclosed small lagoons from the Azov Sea coast at *46° 07′N* and *35° 05′E*. The broad end at a very narrow arm is only possible to develop in shallow water. (Image credit: ©Google Earth 2010).

© 2010 Cnes/Spot Image
Data SIO, NOAA, U.S. Navy, NGA, GEBCO
Image © 2010 DigitalGlobe
Image © 2010 TerraMetrics
©2010 Google

5.35

© 2010 Cnes/Spot Image
Data SIO, NOAA, U.S. Navy, NGA, GEBCO
Image © 2010 DigitalGlobe
Image © 2010 TerraMetrics
©2010 Google

5.36

Image © 2010 TerraMetrics
Data SIO, NOAA, U.S. Navy, NGA, GEBCO
Image © 2010 DigitalGlobe
©2010 Google

5.37

Image © 2010 GeoContent
©2010 Google

5.38

© 2010 Cnes/Spot Image
Image © 2010 TerraMetrics
Data © 2010 MIRC/JHA
©2010 Google

5.39

Image © 2010 TerraMetrics
Data SIO, NOAA, U.S. Navy, NGA, GEBCO
Image © 2010 GeoContent
©2010 Google

5.40

© 2010 Infoterra Ltd & Bluesky
Data SIO, NOAA, U.S. Navy, NGA, GEBCO
Image © 2010 TerraMetrics
Image © 2010 Getmapping plc
©2010 Google

5.41

Image © 2010 European Space Imaging
Data SIO, NOAA, U.S. Navy, NGA, GEBCO
Image © 2010 DigitalGlobe
Image © 2010 GeoEye
©2010 Google

5.42

5.43

Fig. 5.43 A 7 km long tombolo in SE Tasmania, Australia at *43° 16′S* and *147° 19′E.* (Image credit: ©Google Earth 2010).

◂Fig. 5.44 A small tombolo "in statu nascendi" on Elaphonisos Island, southern Greece. (Photo credit: D. Kelletat).

◂Fig. 5.41 Dungeness foreland in southern England at *50° 56′N* and *0° 55′E* is the prototype for a broad headland formed by multiple beach ridges (see also Long et al., 2008). Drift direction has changed many times during the millennia, and more than 500 single beach ridges can be mapped on this headland. Width of image is 14 km. (Image credit: ©Google Earth 2010).

◂Fig. 5.42 The "locus typicus" of all tombolos: Monte Argentario in NW Italy. The central and rudimentary tombolo has been built directly in the wave shadow of the Mt. Argentario Island during a higher sea level in the last interglacial, whereas the two outer tombolos are the result of wave refraction during the Holocene high sea level. Scene is about 18 km wide. (Image credit: ©Google Earth 2010).

Image © 2011 GeoEye
©2010 Google

5.45

Image © 2010 DigitalGlobe
©2010 Google

5.46

◀ Fig. 5.45 Tomboli in statu nascendi behind concrete blocks placed as wave breakers at the Adriatic coast of NE Italy (Image credit: ©Google Earth 2010).

◀ Fig. 5.46 The concrete blocks of the wave breakers can be connected to the mainland beach by b s (Adriatic coast of Italy at *44°* and *12°41′E* with 1.4 km of width). (Image credit: ©Google Earth 2010).

Fig. 5.47 Overwash fan (strong storm or tsunami?) in a barrier of the Rhone delta, southern Mediterranean France. (Photo credit: D. Kelletat).

Fig. 5.48 The barrier at Lesina, Italy, has been washed over by a tsunami leaving sedimentary fans inside the lagoon during the 17th century. The site is at about *41°54′N, 15°28′E* and 5 km wide(Gianfreda et al., 2001). (Image credit: ©Google Earth 2010).

5.4 Barriers, Barrier Islands and Lagoons

Elongated ridges build up by wave action above the high tide level offshore or across the opening of bays or mouths of inlets are called a coastal barrier (Schwartz, 1995). Barriers may consist of sand or gravel (or a combination of both) and usually they are backed by a lagoon or coastal wetlands. Sediment swept over a barrier during storms (or tsunami) are deposited as a washover fan on the inner shore or in the lagoon environment (Fig. 5.47–5.48). The lagoons behind barriers are excellent sediment traps and preserve marine, littoral and terrestrial particles including foraminifera and pollen, and therefore are used to reconstruct climatic and sea level history since their enclosure. The microfossils also can be used to date storm or tsunami events which

Fig. 5.49 The former irregular coastline of Anguilla Island in the north-eastern Caribbean has been straightened by barriers enclosing former bays as modern lagoons. Oblique aerial photograph (Photo credit: A. Scheffers).

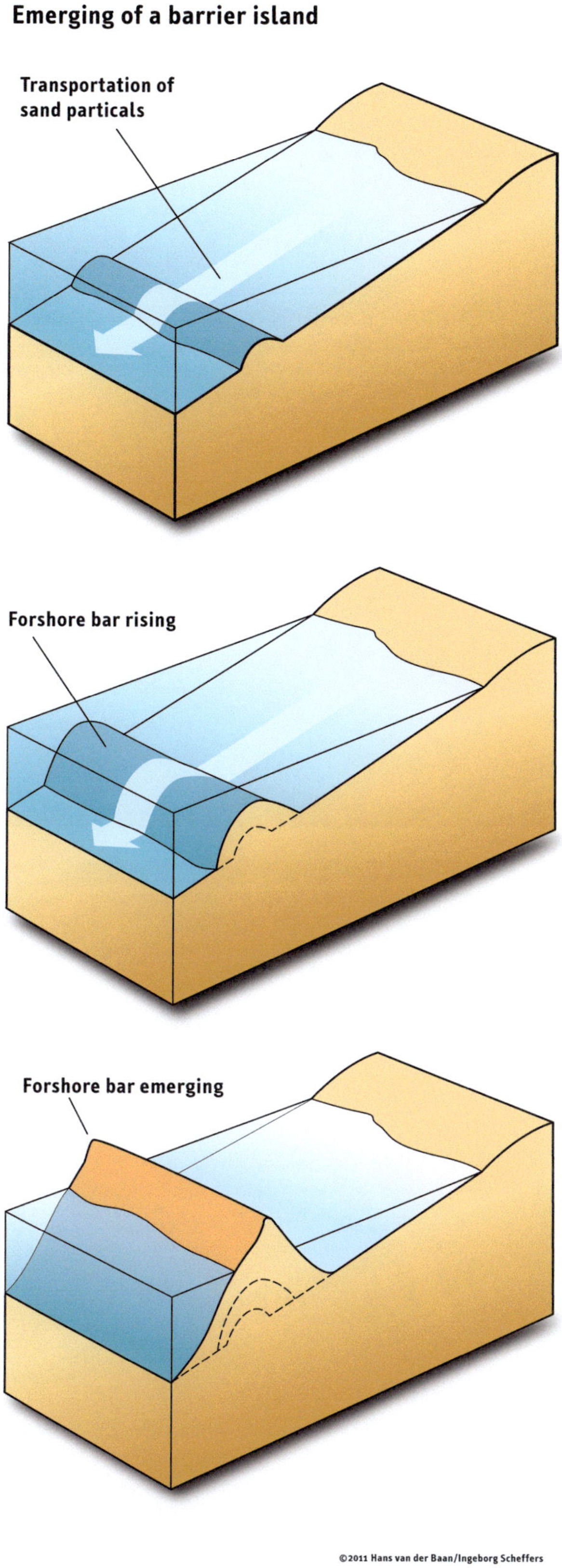

Fig. 5.50 The process of barrier or barrier island development in front of a coastline, where the normal waves touch ground and form offshore sand bars, which later grow to chains of islands.

NEXT PAGES

▶Fig. 5.51 Along the barrier of this lagoon individual spits have developed by inner longshore drift (Brazil at *22° 54′S* and *42° 14′W*, width 40 km). (Image credit: ©Google Earth 2010).

▶Fig. 5.52 Patos lagoon in south-eastern Brazil, with a length of about 250 km, is the largest of the world. (Image credit: ©Google Earth 2010).

▶Fig. 5.53 a) A segmented lagoon in north-eastern Siberia (*69° 49′N* and *175° 46′E*) with a width of 35 km, and (b) enlargement of the north-eastern section, where a part of the outer lagoon is segmented. Width here is only 10 km. (Image credit: ©Google Earth 2010).

▶Fig. 5.54 Lagoons at Nantucket island in NE USA (*41° 16′N* and *70° 02′W*). (Image credit: ©Google Earth 2010).

▶Fig. 5.55 A smooth coastline formed by abrasion of headlands consisting of drift material that is closing the remaining bays by barrier development. Baltic shoreline of Germany and Poland at about *54° 30′N* and *16° 41′E*. length of the shoreline shown is more than 90 km. (Image credit: ©Google Earth 2010).

▶Fig. 5.56 Lagoons with their long axes more or less perpendicular to the coastline and closing barriers are rare closed rias. Martha's Vineyard Island, Massachusetts, north-eastern USA at *41° 21′N* and *70° 36′W*, length about 9 km. (Image credit: ©Google Earth 2010).

▶Fig. 5.57 A barrier has cut a ria coastline from the open sea in the Gulf of Guinea, west Africa, at *5° 17′N* and *4° 31′W*. Width of scene is about 70 km. (Image credit: ©Google Earth 2010).

▶Fig. 5.58 Three barriers and three lagoons show a long history of coastal accretion south of Perth in Western Australia at *32° 52′S* and *115° 41′E*. Width of the image is 10 km. (Image credit: ©Google Earth 2010).

▶Fig. 5.59 A chain of barrier islands in the northern Netherlands about 5 to 10 km in front of the mainland coast, with tidal flats in the shelter of the islands. (Image credit: ©Google Earth 2010).

▶Fig. 5.60 Spiekeroog (nearly 10 km long) as one of the many barrier islands of the German North Sea Coast (Image credit: ©Google Earth 2010).

▶Fig. 5.61 Barrier island chain around Cape Hatteras, east coast of USA. (Image credit: ©Google Earth 2010).

Image © 2010 GeoEye
Data SIO, NOAA, U.S. Navy, NGA, GEBCO
Image © 2010 DigitalGlobe
©2010 Google

5.51

Image © 2010 GeoEye
Data SIO, NOAA, U.S. Navy, NGA, GEBCO
Image © 2010 DigitalGlobe
© 2010 Cnes/Spot Image
©2010 Google

5.52

Image © 2010 TerraMetrics
Image IBCAO
©2010 Google

5.53a

Image © 2010 TerraMetrics
Image IBCAO
©2010 Google

5.53b

Image © 2010 TerraMetrics
Data SIO, NOAA, U.S. Navy, NGA, GEBCO
Image © 2010 DigitalGlobe
©2010 Google

5.54

Image © 2010 DigitalGlobe
© 2010 Cnes/Spot Image
Data SIO, NOAA, U.S. Navy, NGA, GEBCO
Image © 2010 GeoEye
©2010 Google

5.55

Image © 2010 DigitalGlobe
Data SIO, NOAA, U.S. Navy, NGA, GEBCO
©2010 Google

5.56

Image © 2010 TerraMetrics
Image © 2010 GeoEye
Data SIO, NOAA, U.S. Navy, NGA, GEBCO
Image © 2010 DigitalGlobe
©2010 Google

5.57

5.58

Image © 2010 DigitalGlobe
Data SIO, NOAA, U.S. Navy, NGA, GEBCO
©2010 Google

5.59

Image © 2010 TerraMetrics
Data SIO, NOAA, U.S. Navy, NGA, GEBCO
Image © 2010 Aerodata International Surveys
©2010 Google

Image © 2010 TerraMetrics
Data SIO, NOAA, U.S. Navy, NGA, GEBCO
Image NASA
Image © 2010 GeoContent
©2010 Google

5.60

Image © 2011 TerraMetrics
Data SIO, NOAA, U.S. Navy, NGA, GEBCO
Image U.S. Geological Survey
Image © 2011 DigitalGlobe

5.61

have washed over the barriers. In larger lagoons, wind may produce waves with longshore drift that initiate the growth of secondary spits along their inner margins, sometimes separating the lagoons into water bodies of nearly oval form.

Barriers may form by different processes and dependent on the type and availability of sediments, the profile of the coastal slope, tidal range and wave conditions and in addition relative sea level changes (5.49–5.5.58). Many barriers have developed during and after the postglacial sea level transgression by submergence of existing sand ridges and the onshore movement of sea floor sediments. Others have emerged as nearshore bars during relative sea level regressions. Transgressive barriers migrate landward across their back barrier lagoons others remain stable and grow seaward by progradation of beach or dune ridges. Barriers are found along 13% of the world's coastline, in general where the tidal range is small. They may become barrier islands where stronger tidal currents maintain open, transverse gaps.

Barrier islands are long, narrow and unconsolidated sandy islands lying parallel offshore to the main shoreline (Figs. 5.59–5.61). They are separated from the main coast by bays, tidal flats or shallow lagoons. As offshore sandbars build up above the waves, vegetation takes hold and stabilizes the initial islands. Shells, sea grass or algae may accumulate as well, and wind drift may accumulate sand at these obstacles. The process of build-up is accelerated, and finally an island emerges where the first submarine bar developed. Barrier islands are common along tectonically stable, low-lying coasts where tidal ranges do not exceed 4 m and longshore currents are strong (Otvos, 2005). Yet, a new study identifies the world's longest chain of barrier islands along a stretch of the equatorial coast of Brazil, where spring tides reach seven meters and identified a total of 2149 barrier islands worldwide using satellite images and Google Earth imagery, topographical maps and navigational charts (Stutz and Pilkey, 2011). All told, the 2149 barrier islands measure 20,783 kilometers in length, are found along all continents except Antarctica and in all oceans, and make up roughly 10% of Earth's continental shorelines. Seventy-four percent of the islands are found in the northern hemisphere. Prominent barrier islands are found along the coast of the Wadden Sea in Germany, The Netherlands and Denmark. The nation with the most barrier islands is the United States including the coasts of North Carolina, New Jersey and the Texas coast of the Gulf of Mexico. Most of them have been formed during the Holocene sea level high stand of the past 6000 yrs.

▶Fig. 5.62a,b Cheniers, i.e. narrow low ridges of shell fragments in the intertidal zone of the innermost part of the Gulf of Mexico, at *31°22′N* and *114°52′W*, about 5 km wide scene (upper example), and in NW Florida at *30°05′N* and *84°10′W*, 2.3 km wide in the lower image. (Image credit: ©Google Earth 2010).

The morphology and shape of the islands is controlled by a combination of wave and tidal forces. In wave-dominated, microtidal areas, the barrier islands are typically tens of kilometres long, with widely spaced inlets with large flood-tidal deltas and small ebb-tidal deltas. Along wave and tide dominated coast which occur in mesotidal areas, the islands are shorter and wider with abundant inlets and large ebb-tidal deltas and rather small flood-tidal deltas. Like beaches, barrier islands are in dynamic balance (and equilibrium) with the forces that shape them. Some barrier islands (e.g. along the North Carolina coast of the US) regularly receive the full force of destructive hurricanes which can reshape and erode these fragile landforms. Moreover, if this natural balance is disturbed – either by natural changes in wave, current or sea level changes or by anthropogenic influences as real estate development (especially in the temperate zone), these landforms are susceptible to erosion and may in cases even disappear under the sea surface. Over centennial or even decadal timescales, the shorelines of barrier islands can undergo significant changes by forming new inlets, spits or by breaching the existing shoreline. Many homes and other infrastructure are today at risk, but there is little governments or residents can do to prevent these processes from taking their natural course.

5.5 Beach Ridge Systems and Cheniers

Beach ridge plains and foredune plains appear very similar in aerial or satellite imagery and are therefore easily confounded, but form by different depositional agents: Onshore winds initiate foredune development,

© 2010 Cnes/Spot Image
Data SIO, NOAA, U.S. Navy, NGA, GEBCO
Image © 2010 DigitalGlobe
©2010 Google

5.62a

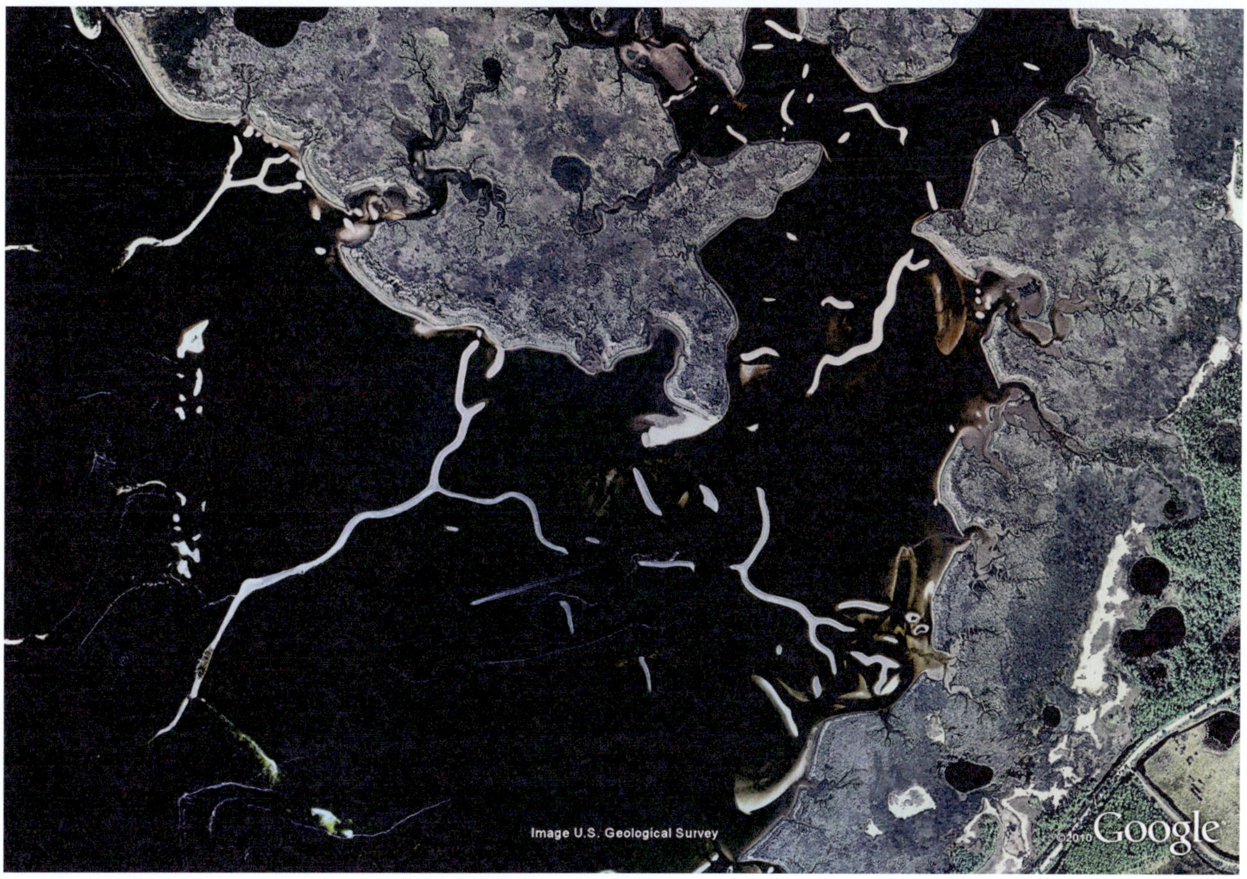
Image U.S. Geological Survey
©2010 Google

5.62b

Fig. 5.63 A chenier composed of pebbles and shell from the east coast of southern Patagonia. (Photo credit: D. Kelletat).

Fig. 5.64 A fresh beach ridge about 0.8 m high made from coral rubble by the 1999 hurricane Lenny in the Caribbean. It shows singular rubble tongues inland from the overwash process, and an avalanching of the rubble on the steep leeward sides. (Photo credit: A. Scheffers).

Fig. 5.65 A beach ridge made from well rounded marble cobbles on the north coast of Ireland. (Photo credit: D. Kelletat).

Fig. 5.66 Single steep beach ridge of quartzite cobbles in north-western Ireland. (Photo credit: A. Scheffers).

Fig. 5.67a,b Portland Bill, a long ridge of pebbles about 3 m above MHW in southern England. (Photo credit: D. Kelletat).

Fig. 5.68 On the Shetland islands in northernmost Scotland pebble and cobble beach ridges may protect formerly open bays. (Photo credit: D. Kelletat).

Fig. 5.69 Coarse debris beach ridges are also typical in paraglacial areas, i.e those which store a lot of sediments of all size from glacial and fluvio-glacial times of the ice ages. The image shows the leeward slope of a 5 m high beach ridge in Galway Bay, west coast of Ireland, deposited more than 1000 years ago (Scheffers et al., 2009, 2010). (Photo credit: A. Scheffers).

whereas beach ridges are deposited by wave action usually during storm events. Beach ridges are composed primarily of sand, pebbles, cobbles (–gravel) or boulders, or a combination of these sediments and form typically at, or above, the normal spring tide high level and can extend as a continuous linear feature for many kilometres along the shore. (Hesp, 2006). Beach ridge development can take place on very short timescales during one storm event or may proceed rather slow over decades depending on event frequency, sediment supply or hydrodynamic conditions. Several studies from cyclone impacts have for example documented that storm wave action can built up ridges of coral rubble more than 3 m high and more than 30 km long during one or two days (Scheffers et al., 2011). A single beach ridge may persist for some time and then be eroded by subsequent storms. Extended beach ridge sequences may form through a series of depositional events over time and contribute to the progradation of the coast (Otvos, 2000; Woodroffe, 2003; Sanjaume & Tolgensbakk, 2009; Taylor & Stone 1996, see also Figs. 5.62–5.76). Once deposited, wind transport may contribute to form superimposed dunes on the beach ridge crests. Most of the ridge systems show some truncations where beach ridge progradation has been terminated by a phase of shoreline erosion, and then followed by renewed ridge growth and thus establishing an often complex pattern of different generations of beach ridges (Figs. 5.70, 5.72 and 5.73).

All beach ridge systems near modern sea level are not older than 7000 years, even those with more than 100 single ridges at one location, when the post-glacial sea level rise reached modern levels. in neotectonic uplifted or glacio-isostatic rising regions, beach ridge systems may reach more than 50 km inland and up to more than 200 m asl., as in the Hudson Bay area of Canada, and the South East region of South Australia.

Cheniers (French chêne for the live oak trees dominating the vegetation on chenier ridges in Louisiana) are similar in morphology and composition to beach ridge system, but differ to the later in that they have been deposited on an alluvial foundation along relatively stable sections in deltas or coastal plains (Figs. 5.62 and 5.63). They are generally emplaced during storm events and can consist of sand, shells or a mixture of both. Cheniers vary in height and width but are usually not more than 2–3 m high and up to 50m wide separated by relatively wide intertidal mud flats. Extended chenier plains occur along the coasts of Louisiana (US) and French Guiana (South America), and are also typical in Northern Australia along the Van Diemen Gulf.

Fig. 5.70a,b Beach ridge systems composed of coarse coral rubble in the Abrolhos archipelago off Western Australia. (Photo credit: A. Scheffers).

▶Fig. 5.71 Excellent preservation of beach ridges from strong wave impact along the west coast of Mexico at about *22°N* and *105°35′W*. Width of image is about 20 km. (Image credit: ©Google Earth 2010).

▶Fig. 5.72 A beach ridge system from Brazil at *17°40′S* and *39°13′W* with a width of 20 km. (Image credit: ©Google Earth 2010).

Data SIO, NOAA, U.S. Navy, NGA, GEBCO
© 2010 Cnes/Spot Image
Image © 2010 DigitalGlobe
©2010 Google

5.71

Data SIO, NOAA, U.S. Navy, NGA, GEBCO
Image © 2010 DigitalGlobe
Image © 2010 GeoEye
©2010 Google

5.72

Fig. 5.73 St. Vincent Island in NW Florida (USA) is a result of beach ridge accretion over several thousand years. Site is at *29°39′N* and *85°08′W* with a width of 13 km. (Image credit: ©Google Earth 2010).

Fig. 5.74 A series of beach ridges closes this lagoon from the open sea in easternmost Russia at *60°N* and *170°09′E*. Width of scene is about 25 km. (Image credit: ©Google Earth 2010).

Fig. 5.75 A beach ridge system truncated by the modern shoreline in a barrier of the west coast of the Black Sea at *44°30′N* and *28°46′E*. Width of scene is 25 km. (Image credit: ©Google Earth 2010).

Fig. 5.76a,b,c Boulder ridges composed of angular blocks up to 80 tons, 150 m from the cliff edge and at +15 m asl on Inishmaan Island, Aran Islands, central west coast of Ireland. (Photo credit: D. Kelletat).

5.6 Coastal Dunes

Dunes are composed of windblown sand and thus form only in settings that have a ready supply of loose sand. Along shorelines the principal sources of sand are streams and rivers carrying sand to the ocean, the gradual weathering of rock formations or cliffs exposed along the coast or fragments of shell, coral and other skeletal fragments transported onshore by waves and currents. Most dunes are found in dry climates as wind cannot pick up wet materials easily because they are too cohesive, but dunes along a coast are an exception to this general rule: At coasts, sand is so abundant and dries very quickly in strong onshore wind that dunes can form even in humid climates (Woodroffe, 2003; Davidson-Arnott, 2010; Short & Woodroffe, 2009; see also Figs. 5.77–5.84).

Onshore winds erode dry sand from the steepening face of the beach and deposit the windblown sediments towards the upper beach where foredunes gradually form. Wind as a transport agent can move large amounts of sediment. Imagine that within one day a strong wind of 48 km/hour can transport half a ton of sediment from a 1m wide strip across a sand dune's surface. The sand particles are moving by saltation, a process whereby individual sand grains are carried by the wind close to the surface in a series of short hops. Wind action effectively sorts the beach material and sand deposited in dunes is essentially of one size (diameter ranging from 0.15 to 0.30 mm). Once foredunes are formed, vegetation plays a dominant role in determining their stability, size and shape. Grass vegetation such as Spinifex sp. has a particular ability to produce upright stems and new roots in response to constant sand covering.

►Fig. 5.78 A field of parabolic sand deposits at the California coastline at about *35° N* and *120° 37′ W.* (Image credit: ©Google Earth 2010).

►Fig. 5.79 Somalia's Red Sea coast at *13° 46′ N* and *42° E* exhibits some fine parabolic sand deposits. Length of coastline shown is about 25 km. (Image credit: ©Google Earth 2010).

Fig. 5.77 A beach with different generations of coastal dunes during low water (Ireland). (Photo credit: A. Scheffers).

Image © 2010 DigitalGlobe
Data SIO, NOAA, U.S. Navy, NGA, GEBCO
Data CSUMB SFML, CA OPC
©2010 Google

5.78

© 2010 Cnes/Spot Image
Image NASA
Data SIO, NOAA, U.S. Navy, NGA, GEBCO
©2010 Google

5.79

Image © 2010 DigitalGlobe
Data SIO, NOAA, U.S. Navy, NGA, GEBCO
Image © 2010 TerraMetrics
Google

5.80

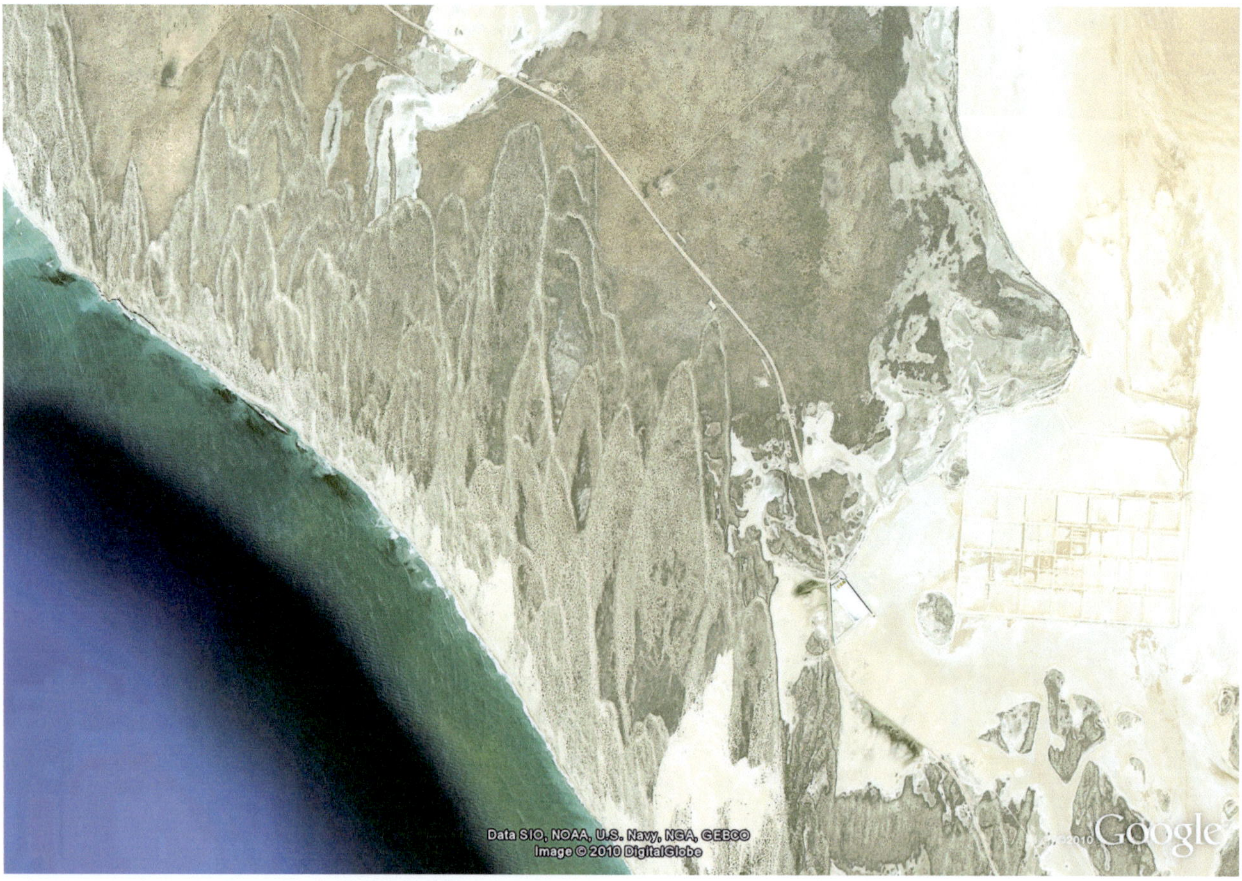
Data SIO, NOAA, U.S. Navy, NGA, GEBCO
Image © 2010 DigitalGlobe
Google

5.81

Fig. 5.82 Old vegetated parabolic sand dunes along a narrow beach on the south coast of Queensland, Australia, at *22° 51′S* and *150° 45′E*. Width of image 18 km. (Image credit: ©Google Earth 2010).

Successive stages of plant growth and sand deposition result in increased height and width of the new dune generation. Over time, a series of dunes parallel to the shoreline may develop. Geologists call these long sand dune ridges oriented at right angles to the wind direction transverse dunes. They typically form in beach environments with strong onshore winds but are also common arid desert regions where there is abundant sand and vegetation is absent. When the vegetation cover is damaged either naturally by drought, fire or cyclones or artificially by construction, grazing or motor vehicle / pedestrian tracks, the wind can initiate blowout formation. Also, natural gaps in a foredune system can be widened and transformed into blowouts by onshore winds. Unless the blowouts are stabilized by vegetation they can increase in size and migrate inland in response the onshore wind regime. In this way, blowouts can develop into parabolic dunes which have a slip face that is convex downwind (Fig. 5.79, 5.81–5.83) Parabolic dunes often disrupt a system of transverse dunes or beach ridge systems. Crescent-shaped barchan dunes (Figs. 5.80 and 5.84) are almost the reverse of parabolic dunes with horns of the crescent point downwind. Barchan dunes are commonly found in desert environments and are a rare occurrence along coastlines, but an exquisite example can be seen along the semi-humid tropical coast of Brazil (Fig. 5.80). The literature on coastal dunes is very extensive, in particular in combination with foredune and barrier development. Many of the parabolic dune forms presented in the satellite images are old and not active.

◀ Fig. 5.80 Well nourished beaches in northern Brazil, south of the Amazon delta at *2° 29′S* and *43° 09′W*) provide an abundance of sand to initiate dune formation by onshore winds. The seaward part shows indifferent and parabolic forms, the inland section barchans, which normally are typical for deserts, but here occur in a humid tropical environment. Width of image, 32 km. (Image credit: ©Google Earth 2010).

◀ Fig. 5.81 Perfect parabolic forms of sand dunes along the Western Australian coastline in the vicinity of Shark Bay at *24° 32′S* and *113° 28′E*. Width of image is 10 km. (Image credit: ©Google Earth 2010).

Fig. 5.83 Parabolic sand dunes stabilized by vegetation on the south coast of Western Australia, west of Albany. Width of scene about 15 km. (Image credit: ©Google Earth 2010).

Fig. 5.84 On the outer barrier of the Rhone delta in southern Mediterranean France barchan dunes may develop in dry summers. Oblique aerial photograph (Photo credit: D. Kelletat).

5.7 Marine Deltas

Deltas are coastal landforms that evolve at the mouth of rivers and streams which deposit sediments to the coastline. When waves, tides or currents are not capable to erode all new sediment carried to the sea by streams, the sediments builds outward to form a marine delta. Deltas can be found on all continents and in all climate zones. The term delta was coined by Herodotus in the 5th century BC who noted a geometric similarity between the roughly triangular shape of sediments at the mouth of the Nile River and the Greek letter Δ. Their shapes and sizes are very diverse and variable, but in general reflect a balance reached between sedimentation and erosion along a coast (Figs. 5.85–5.98). Globally, modern delta systems are geological young features that have started to form when the rate of the postglacial sea level rise dropped below a critical threshold to allow delta progradation and a relatively synchronous development of the world's deltas.

Important factors for delta development and the specific morphology of a delta include fluvial characteristics such as the nature of the lower river course and the availability and grain size of sediments delivered to the coast and marine processes: here, tidal range, wave climate, coastal currents and water depth are the primary influences. Coastal researchers often use a classification scheme developed by Galloway (1975) on the basis of dominate processes (fluvial/wave/tide-dominated) and the resulting morphology.

In low energy environments with minimal wave energy and small tidal ranges, the river system is dominant in the morphological characteristics of a delta. In cases these deltas may protrude quickly towards the sea as is exemplified by the modern Mississippi delta, representative of the so called elongated "Bird Foot" delta types (Figs. 5.97 and 5.98). In contrast, cuspate deltas (Figs. 5.85 and 5.88) are dominated by the

Fig. 5.85 The Nile delta with its separated delta mouths of the Damietta and Rosetta rivers. These parts of the delta are now under significant erosion, because the Assuan High Dam of the Nile River retains most of the sediments which formerly fed the delta. Width of image is nearly 400 km! (Image credit: ©Google Earth 2010).

Data SIO, NOAA, U.S. Navy, NGA, GEBCO
© 2010 Cnes/Spot Image
Image © 2010 DigitalGlobe
Image © 2010 TerraMetrics
©2010 Google

5.86

Data SIO, NOAA, U.S. Navy, NGA, GEBCO
Image © 2010 TerraMetrics
©2010 Google

5.87

Fig. 5.88 This cuspate delta at the north coast of the Yucatan Peninsula of Mexico shows continuous progradation with beach ridge sequences at *18° 32′N* and *92° 39′W*. (Image credit: ©Google Earth 2010).

nearshore wave climate and often destructional with well-developed strand plains as typical for the Sao Francisco Delta in Brazil. Lobate deltas (Figs. 5.86, 5.90–5.93), exemplified by the Niger Delta in Africa, are intermittent between constructional and destructional forms and often possess a smooth coastal outline. Deltas shaped predominantly by tides may develop inlets and distributary channels that are straight and orientated parallel to the tidal movements as is the case in the Ganges-Brahmaputra or the Amazon deltas. "Truncated" deltas also occur (Fig. 5.89), with almost no advance seawards of a strait shoreline.

◄ Fig. 5.86 Part of the Shatt-el-Arab, the joint delta of Euphrates and Tigris rivers in the Persian Gulf at *30° 05′N* and *48° 48′E*, 24 km wide. In this desert environment dry intertidal flats are typical. (Image credit: ©Google Earth 2010).

◄ Fig. 5.87 A simple triangular delta form at the west coast of Africa (Gulf of Guinea) at *3° 34′N* and *9° 41′W*. Width of image is 32 km. (Image credit: ©Google Earth 2010).

Some complex deltas have a long and complicated history. As a delta grows in one direction for some hundreds or thousands of years, it then breaks out to form a new distributary channel and begins to grow into the sea in another direction. For example, the Mississippi River has built a series of overlapping deltas of different Holocene ages which over the past 6000 yrs shifted both to the east and the west.

The world's great deltas and associated wetlands hold rich, fertile soils constituting prime agricultural lands with similar abundance of food resources in the adjacent marine areas. Today, they are home to millions of people as is true for Egypt's Nile Delta, China's Yangtze and Huang He deltas, the Mississippi delta on the Gulf Coast of the United States or the Ganges Delta of Bangladesh. Delta environments with their rich soils and the abundance of game and fish food resources allowed also for the establishment and expansion of prehistoric cultures and served as the cultural hub for ancient civilisations. Our modern technologically driven society explored

Image © 2010 TerraMetrics
Data SIO, NOAA, U.S. Navy, NGA, GEBCO
Image © 2010 DigitalGlobe
©2010 Google

5.89

© 2010 Cnes/Spot Image
Image © 2010 TerraMetrics
Data SIO, NOAA, U.S. Navy, NGA, GEBCO
Image © 2010 DigitalGlobe
©2010 Google

5.90

and exploited ancient deltas as they often contain a wealth of fossil fuels such as hydrocarbons or coals. In many areas of the world, deltas have suffered from human activities: Since the 1930's engineers have built flood controlling structures and dams along the river course and decreased the amount of sediment delivered to the delta thereby also preventing the small but frequent floods that nourish the delta wetlands in a natural state. Worldwide over 41,000 large dams are in operation in addition there are also many smaller dams and water reservoirs. Together, they block 14% of the total global river flow, as well as enormous volumes of sediment (Syvitskivet al., 2009). This anthropogenic induced change adds to the process of subsidence in deltas: As we have seen, deltas grow with the addition of sediment, and they sink as the sediment becomes compacted and the crust subsides under the weight of the sediment load. Today, entire cities such as Venice, built on a delta in Northern Italy or Bangkok, Thailand's sprawling capital of more than 10 million people (once dubbed "Venice of the East") built on the banks of the Chao Phraya River delta are sinking. The situation is worsened with groundwater, oil or natural gas being pumped out (both legally and illegally) to supply residential and industrial demands. A recent scientific study of 33 major river deltas in the world revealed that 24 of them are sinking (Syvitskivet al., 2009). The prospect of a watery future for these cities is amplified by the recent sea level rise. The future of deltas will depend on our wisdom and ability to balance environmental management strategies and conservation versus economic development.

◀Fig. 5.89 The beach ridges of this truncated delta in Mexico at *18°37′N* and *92°27′W* show that it formerly reached farther into the sea (Central America). The reason for delta erosion maybe subsidence, stronger waves/storms, lack of sediments, or a combination of all these factors. Width of image is 25 km. (Image credit: ©Google Earth 2010).

◀Fig. 5.90 This debris fan at the east coast of Baja California (Mexico) is also a delta. *30°15′N* and *114°39′W*, image 12 km wide. (Image credit: ©Google Earth 2010).

NEXT PAGES

▶Fig. 5.91 A rounded delta at the south coast of Iran in the sheltered Persian Gulf at *26°49′N* and *55°30′E*. (Image credit: ©Google Earth 2010).

▶Fig. 5.92 116 The complex Danube delta in the Black Sea. North-south extension is about 150 km. (Image credit: ©Google Earth 2010).

▶Fig. 5.93 Arctic deltas a) at the northwest coast of Alaska at about *63°10′N* and *164°10′W* with a width of about 40 km. The complex pattern of channels in the delta result from an ever changing strong run- off in the short time of melt water discharge, which may last only some weeks per year. Inspite of a rather dry climate, impermeable permafrost makes a swampy environment with a lot of ponds and open water arms. (Image credit: ©Google Earth 2010). (b) Arctic delta. Northern Russia at *73°24′N* and *127°14′E* with a width of 50 km. (Image credit: ©Google Earth 2010).

▶Fig. 5.94 A Russian delta into the northern Black Sea at *46°31′N* and *32°22′E*, image is 25 km wide. This is an example that deltas may even appear in the inner parts of bays, like the la Plata delta in the "Rio de la Plata" in northern Argentina. (Image credit: ©Google Earth 2010).

▶Fig. 5.95 Typical meandering tidal creeks in mangrove forests in tropical deltas of very flat lowlands and a shallow sea (Ganges-Brahmaputra, India, at *21°54′N* and *89°45′E*, 30 km wide). (Image credit: ©Google Earth 2010).

Image © 2010 DigitalGlobe
© 2010 Cnes/Spot Image
Data SIO, NOAA, U.S. Navy, NGA, GEBCO
Image © 2010 GeoEye
©2010 Google

5.91

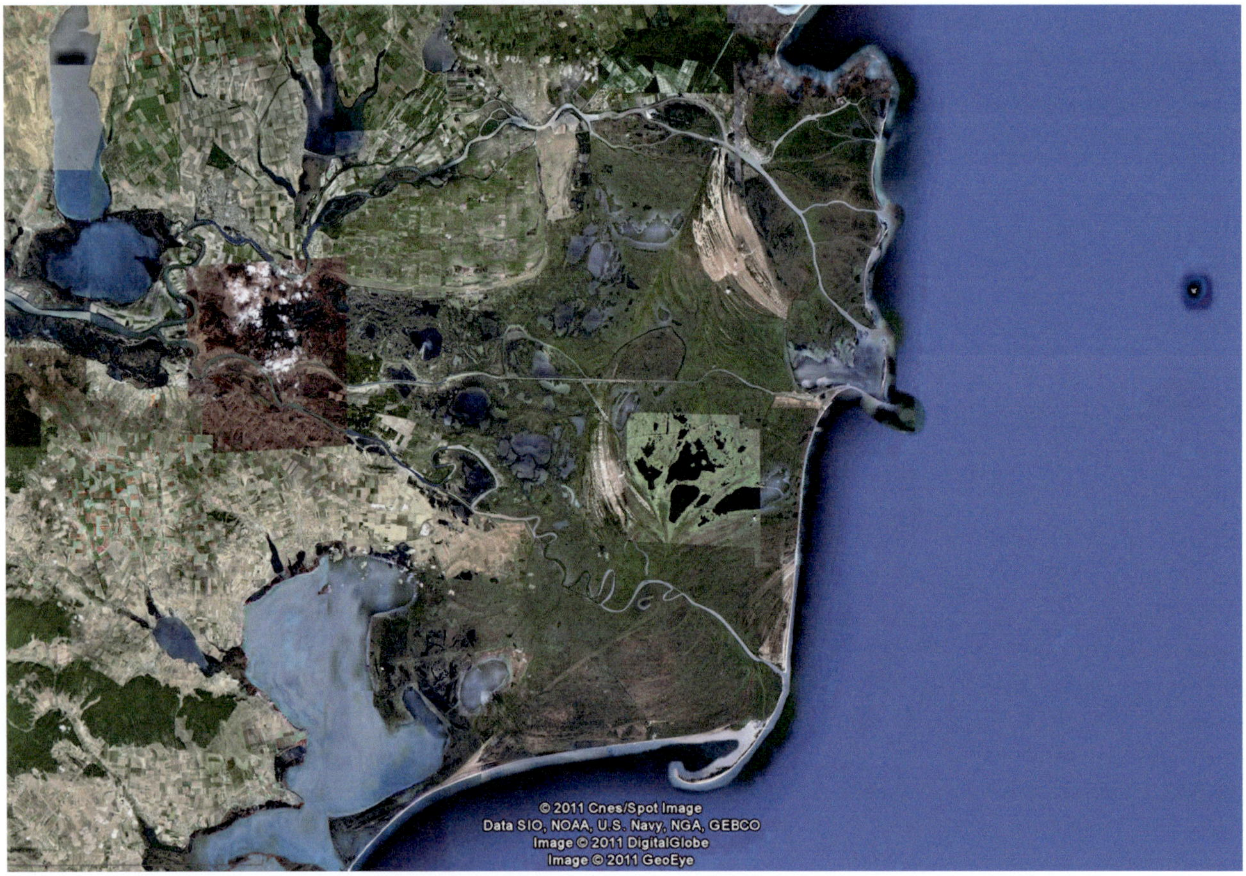
© 2011 Cnes/Spot Image
Data SIO, NOAA, U.S. Navy, NGA, GEBCO
Image © 2011 DigitalGlobe
Image © 2011 GeoEye

5.92

Image © 2010 TerraMetrics
Image IBCAO
Image © 2010 DigitalGlobe
Data SIO, NOAA, U.S. Navy, NGA, GEBCO
Google

5.93a

Image © 2010 TerraMetrics
Image IBCAO
©2010 Google

5.93b

Image © 2010 European Space Imaging
Image © 2010 TerraMetrics
©2010 Google

5.94

Image © 2010 TerraMetrics
Image © 2010 GeoEye
©2010 Google

5.95

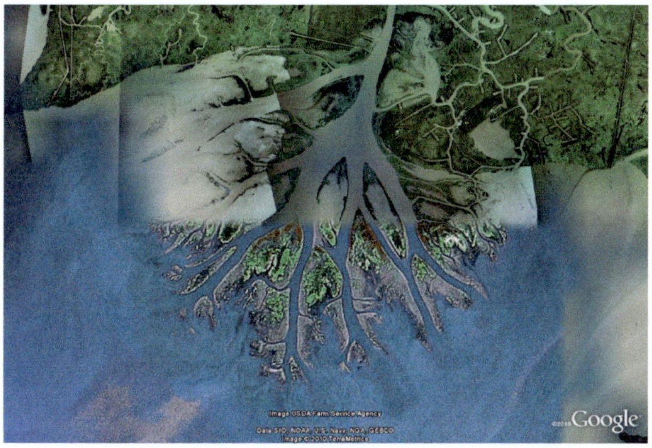

Fig. 5.96 This 10 km wide delta shows that the forces of the sea are not dominant and delta sedimentation and progradation is not transformed by wave action. West of the Mississippi delta at *29° 30′N* and *91° 26′W.* (Image credit: ©Google Earth 2010).

Fig. 5.97 Deltas sheltered from strong waves and currents show the fluvial regime dominating the marine one. Example from Venezuela at *10° 10′N* and *75° 32′W* with a width of the image of 9 km. (Image credit: ©Google Earth 2010).

Fig. 5.98 Main delta mouths of the Mississippi birdfoot delta *(29° 09′N* and *89° 03′W),* exhibiting natural levees along the outer river courses as narrow elevated banks. Width of the scene is about 35 km. (Photo credit: ©USGS).

References

Davidson-Arnott R (2010) Introduction to Coastal Processes and Geomorphology. Cambridge University Press, Cambridge.

Galloway WE (1975) Process framework for describing the morphologic and stratigraphic evolution of deltaic depositional systems. In: Broussard ML (ed.) Deltas: Models for Exploration. Houston, Houston Geological Society: 87–96.

Gianfreda F, Mastronuzzi G, Sanso P (2001) Impact of historical tsunamis on a sandy coastal barrier: an example from the northern Gargano coast, southern Italy. Natural Hazard and Earth System Science 1: 213–219.

Hesp PA (2006) Sand Beach Ridges: Definitions and Re-Definitions. Journal of Coastal Research Special Issue 39:72–75.

Kelletat D, Scheffers A (2005) Europe, Coastal Geomorphology. In: Schwartz M (ed) Encyclopedia of Coastal Science. Springer, Dordrecht.

Komar, PD (1998) Beach processes and sedimentation, 2nd edn. Upper Saddle River, NJ; Prentice-Hall.

Long A, Waller MP, Plater AJ (eds, 2008) Dungeness and Romney Marsh: Barrier Dynamics and Marshland Evolution. Oxbow Books Ltd, Oxford.

Otvos EG (2000) Beach ridges—definitions and significance. Geomorphology 32:83–108.

Otvos EG (2005) Coastal Barriers, Gulf of Mexico: Holocene Evolution and Chronology. Journal of Coastal Research Special Issue 42: 141–163.

Schwartz ML (Ed., 2005) Encyclopedia of Coastal Science. Springer, Dordrecht.

Sanjaume E, Tolgensbakk J (2009) Beach ridges from the Varanger Peninsula (Arctic Norwegian coast): Charactarisαtics and significance. Geomorphology 104 (1–2):82–92.

Scheffers A, Engel M, Scheffers S, Squire P, Kelletat D (2011) Beach Ridge Systems Archives for Holocene Coastal Events? Progress in Physical Geography. doi: 10.1177/0309133311419549.

Short AD (2006) Australian Beach Systems – Nature and Distribution. Journal of Coastal Research 22 (1):11–27.

Short AD, Woodroffe CD (2009) The Coast of Australia. Cambridge University Press, Cambridge.

Stutz ML and Pilkey OH (2011) Open-Ocean Barrier Islands: Global Influence of Climatic, Oceanographic, and Depositional Settings. Journal of Coastal Research, 27 (2): 207–222.

Syvitski JPM, Kettner AJ, Overeem I, Hutton EWH. Hannon MT, Brakenridge GR, Day J, Vrsmarty C, Saito Y, Giosan L, Nicholls RJ (2009) Sinking Deltas due to Human Activities. Nature Geoscience, 2, 681–686.

Taylor M, Stone GW (1996) Beach ridges: a review. Journal of Coastal Research 12:612–621.

Woodroffe CD (2003) Coasts: Form, Process and Evolution. Cambridge University Press, Cambridge.

6

Coasts Dominated by Organisms

ABSTRACT Organisms, either plants or animals, at the interface of land and sea, form among the most productive and biologically complex ecosystems on Earth. Coral reefs, mangrove forests or sea grass meadows are important not only from an ecological point of view, but provide tens of millions of people in the tropics and subtropics with food, timber, fuel or medicine. The most prominent organisms that form vast 3-dimensional coastal structures are reef-building corals. They occur between approx. 30°N and S in all oceans, where the water is clear and over 18°C in the coldest month. Coral reefs occupy less than one tenth of one percent of the world ocean surface, about half the area of France, yet they provide a home for twenty-five percent of all marine species. Sea grass and mangrove forests are land builders par excellence by trapping sediments. Intact, they serve as natural breakwaters and dissipate the energy of waves along the coast. Despite their strategic importance, these ecosystems are under threat worldwide: They are destructed for salt pans, aquaculture enterprises, infrastructure and housing developments and suffer from indirect stresses such as climate warming, chemical pollution, sediment overload and disruption of a water and salinity balance, just to name a few. Moreover, many marine organisms also provide important insight in past environments, sea levels or tidal ranges, which might hold the answer to preserving ecosystems in a rapidly changing climate.

6.1 Marine Plants – Algae and Seagrass

Plant life is widespread and very diverse in coastal waters including mangrove forests, saltmarsh vegetation, seagrass meadows and algal communities. These environments provide habitat for marine mammals, fish and shellfish and nursery areas to the larger ocean, and performing important physical functions of filtering coastal waters, dissipating wave energy and anchoring sediments. Macroscopic algae in the ocean, such as *Sargassum sp.* and kelp are commonly known as seaweeds and create kelp forests. The non-algae plants that survive in the sea are often found in shallow waters, such as the seagrasses or mangrove forests in the intertidal zone, which have adapted to the high salinity of the ocean environment.

Seagrasses, a group of about sixty species of underwater marine flowering plants, grow in the shallow marine and estuary environments of all the world's continents except Antarctica (Figs. 6.1–6.3; Green and Short, 2003). They thrive in shallow coastal waters with low wave energy and sandy or muddy bottoms and have the same basic structure as land plants. Seagrasses produce flowers; they may have strap-like blades as leaves (as the

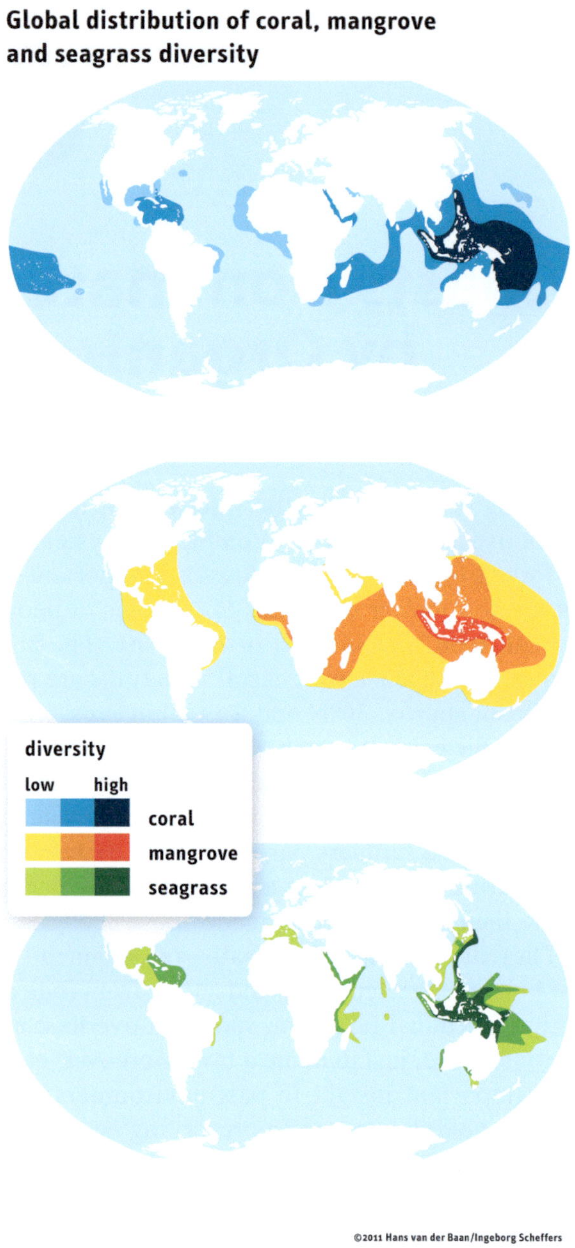

Fig. 6.1 These world maps show the importance of organics in coastal waters (Source: UNEP-WCMC, 2001).

Fig. 6.2 Dark meadows of sea grass in the Bahamas. (Photo credit: S. Scheffers).

Fig. 6.4 Storms form these "sea-balls" from the stronger parts of sea grass leaves (Sardinia, Italy). (Photo credit: D. Kelletat).

►Fig. 6.3 Sea grass patterns in Shark Bay, Western Australia, at *25° 58′S* and *113° 47′E*. Scene is about 13 km wide. (Photo credit: ©Google Earth 2010). *When seen from a bird's eye perspective, seagrass in Shark Bay often takes on a banded appearance. Seagrass grows in bands that form perpendicular to the water flow.*

►Fig. 6.5 The dark colours in the intertidal zone represent dense stands of *Fucus* and Laminaria (brown algae, France, at *48° 40′N* and *4° 17′W*, with width of scene of 5 km). (Photo credit: ©Google Earth 2010).

eelgrass *Zostera caulescens* in the Sea of Japan, at more than 4 m long) or oval leaves like the tiny sea vine (e.g. *Halophila decipiens*) and a root system. Seagrasses are the primary food of manatees, dugongs and green sea turtles, all threatened and charismatic species of great public interest, and are a critical habitat for thousands of other animal and plant species. The root and rhizome system of seagrass catches and stabilizes sediments and protects the coastline from erosion by currents or pounding waves. Storms can deposit huge amounts of seagrass onshore, sometimes transformed by wave action into thousands of "sea balls" (Fig. 6.04).

Many species of macro-algae (green, brown, red) grow close to the coast and particularly on hard rock that they need for attachment (Figs. 6.05–6.11). The most significant algal zones along the shorelines of the world are located in cool waters, i.e. notably in upwelling areas of cold water near the coastline (e.g. countries along the west coast of continents such as California, Chile, Peru, Namibia). Giant kelp such as *Macrocystis sp.* and *Nereocystis sp.*

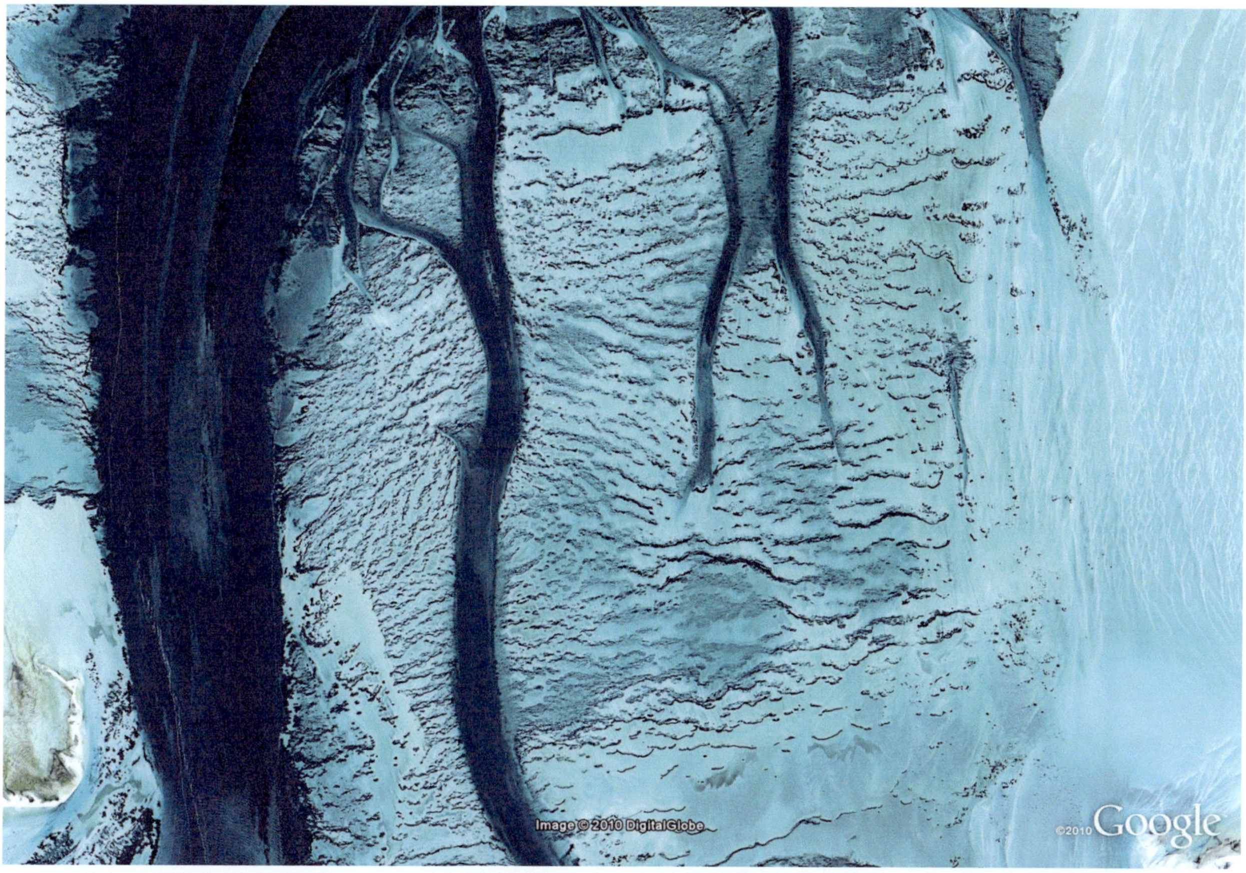
Image © 2010 DigitalGlobe
©2010 Google

6.3

Image © 2010 DigitalGlobe
© 2010 Cnes/Spot Image
Image © 2010 IGN-France
Data SIO, NOAA, U.S. Navy, NGA, GEBCO
©2010 Google

6.5

Fig. 6.6 a, b A belt of brown algae surrounds all rocky islands in the Outer Hebrides, Orkneys and Shetlands of Scotland. (Photo credit: D. Kelletat).

form underwater forests with dense canopies up to 35 meters above the seabed and inhabit many cool water environments. Kelp usually grows on subtidal rocky reefs although some species are able to grow on smaller scattered rocks. In general they grow on reefs in waters to a maximum of around 30 meters depth, although most are found in shallower waters. Unlike land plants seaweeds do not gain their water or nutrients such as nitrogen and potassium from the ground but instead absorb both directly from the water through surface of the plant. Marine algae vary in their needs for nutrients and some species such as aforementioned giant kelps are unable to tolerate low nutrient levels. Thus these species are usually found in areas with high nutrient levels (upwelling areas or in places such as near seal colonies where nutrient levels are raised by for example seal excrements).

Algae use a variety of strategies to remain attached to the rocks, even under the enormous forces of waves crashing down or subjected to strong currents. These attachment strategies include the use of powerful adhesives, which may have potential for human use, and physical structures that take advantage of small imperfections in the rock surface. Brown algae (e.g. kelp) attach themselves to solid structures such as rock, and extend their leaves into the waters above them as they reach towards the sunlight. Strong structures at the bases of kelp plants called holdfasts hold the kelp firmly attached to rocks, so that even when currents and waves are at their strongest, kelp is in no danger of being swept away. In addition to these holdfasts, kelps also have stem like struc-

Fig. 6.7 A "kelp forest" of giant *Macrocystis pyrifera* in California. (Photo credit: NOAA, www.photolib.noaa.gov).

Fig. 6.8 The intertidal giant kelp *Durvillea antarctica* in southern New Zealand with large holdfast, diameters up to about 0.4 m. (Photo credit: D. Kelletat).

Fig. 6.9 Floating parts of Durvillea antarctica of southern New Zealand. The single "leaves" may reach a length of more than 5m. (Photo credit: D. Kelletat).

Fig. 6.10 Another aspect of floating Durvillea antarctica. (Photo credit: D. Kelletat).

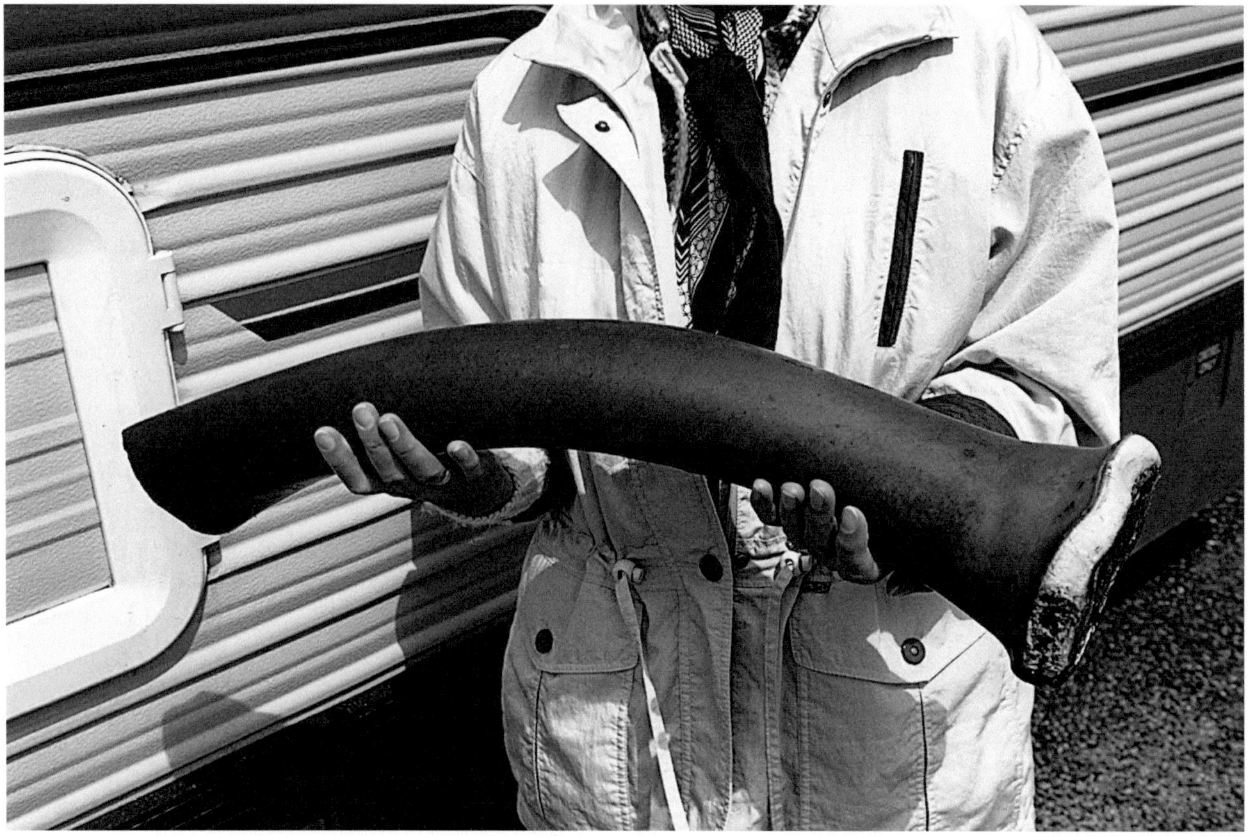

Fig. 6.11 This thick "stem" of Durvillea antarctica shows the strength of these algae in the surf belt. Largest single plants may grow up to 100 kg in several months. (Photo credit: D. Kelletat).

tures for support of blades, known as stipes. In some species gas filled vesicles, or bladders, provide uplift for the blades and stipes and keep the plants up in the water column allowing for better access to light.

These marine plants also withstand partial exposure to the air and direct sunlight during low tide. Another challenge for larger marine plants is to overcome the properties of seawater that result in light being selectively absorbed with depth. As light penetrates water, certain colors are absorbed. The blades of kelp are similar to tree leaves in that they contain the pigments that allow them to absorb sunlight and convert it into carbohydrates through photosynthesis. The different colors of algae are due to the presence of specific pigments that allow them to absorb different wavelengths in the light spectrum, which allows utilizing different niches (i.e. here: depths).

The very tough and resistant algae such as Durvillea antarctica protect rocky shores from wave erosion and reduce evaporation during low tides. All these large macroalgae have strong holdfasts fixed to the substrate in the rocky intertidal zones (Durvillea sp.) or on rocks or boulders in the foreshore. As mentioned before, these algae have carbon dioxide (gas) filled bladders that make them buoyant and, when during very strong wave activity, they might be pulled from their anchoring places and drift ashore, sometimes with pieces of substrate attached to their holdfasts.

Furthermore, kelp forests are important marine habitats and provide shelter and food for large populations of fish, crustaceans, mollusks, echinoderms and marine mammals like seals or sea otters (Steneck et al., 2002). In temperate regions (Scotland, Norway, eastern USA and Canada) algal species such as Fucus sp. and Laminaria sp. are common.

The most common plant genus near the high tide level is Salicornia sp., whereas Spartina sp. grass grows in positions just higher than the tidal zone. Both contribute significantly in stabilizing soil and catching suspended sediments; therefore these plants promote sediment accumulation in shallow coastal waters. These plants can also be found in the topographically higher situated marshlands, which are normally only flooded dur-

Fig. 6.12 In some latitudes driftwood may form extensive and persistent deposits at the coastline (example from Washington State, USA). (Photo credit: D. Kelletat).

ing extreme storm tides, i.e. once a year. In cold regions, marshlands may be affected by deposition and movement of drift ice, which kills off patches of grass, leading to numerous water ponds in the marsh. Another destructive factor in higher latitude marshlands is the bi-annual visit of tens of thousands of migrating birds, notably Canada geese. When these birds feed on the grass roots, the flocks "erode" some centimeters of marshland within a few weeks.

6.2 Marine Plants – Mangroves

Mangroves are plants with more than 110 different species of which only 54 species constitute the true mangroves, i.e. species that occur almost exclusively in mangrove habitats. Mangroves grow in the tropics and subtropics in saline intertidal coastal habitats, such as estuaries (Duke, 2006; Kelletat, 1995 and Figs. 6.13–6.17). Mangrove ecosystems are highly productive areas and very important from economic and ecological points of view. Where these habitats occur, millions of people depend on a variety of mangrove forest products, such as dyes, wood, medicines, livestock nourishment etc. Mangroves host a wide variety of organisms, including many, nowadays, endangered species. These habitats serve as a nursery and feeding ground to (reef) fish, and invertebrates such as crustaceans and mollusks. They also maintain water quality by filtering pollutants and increase land area by trapping sediments. Mangroves also help prevent land erosion by stabilizing substrate and protecting the coast from storms surges and waves.

Mangroves and mangrove forest communities are unique and physiologically adapted to overcome the problems of anoxia, high salinity and frequent tidal inundation. Mangroves occupy the upper half of the tidal range, standing in water up to their leaves during high tide, but are exposed during low tide. The vast majority of mangrove species can be found in the wet tropics around the equator and with their center of diversity in Indonesia (Fig. 6.17); only two species are common in the higher latitudes around 35°N and S;

Fig. 6.13 Pneumatophores (roots for breathing and sometimes exuding excess salt) of *Avicennia* sp. in Florida, USA. (Photo credit: D. Kelletat).

Fig. 6.14 Stilt roots of the *Rhizophora* species, northern Australia. (Photo credit: S. Scheffers).

Fig. 6.15 *Sonneratia* of the inner tropics has strong and high pneumatophores, often with barnacles and oysters attached. Example from the Palau islands, Micronesia. (Photo credit: D. Kelletat).

Fig. 6.16 Knee-like roots of a *Bruguiera* species from Queensland, Australia. (Photo credit: S. Scheffers).

Distribution of mangroves

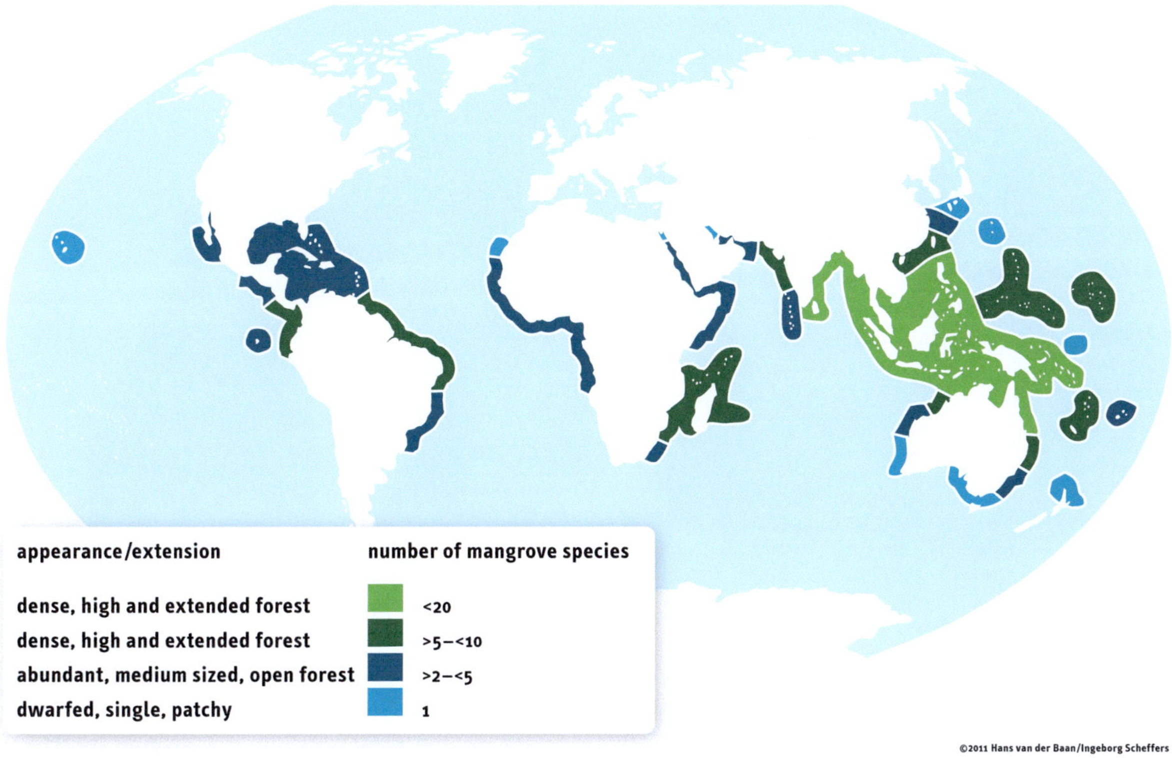

Fig. 6.17 The distribution of mangrove species, varieties, and growth forms and densities are variable in the tropical oceans, and the centre of species development can be clearly identified in the Malay-Indonesian archipelago and northern Australia (Kelletat, 1999).

Avicennia sp. and *Rhizophora sp.* Mangroves have developed special root systems to fix themselves in muddy substrate. These root systems come in many shapes, e.g. as vertical pins growing from horizontal hidden main roots (*Avicennia sp.*, Fig. 6.13), knee-like stilts (*Rhizophora sp.*, Fig. 6.14), thick stems growing from the ground and covered by oysters and barnacles (*Sonneratia sp.*, Fig. 6.15), or knee-like roots (*Bruguiera sp.*, Fig. 6.16). Between these roots a protected environment is created during high tide, which functions as nursery for young fish, crabs and other organisms. Mangroves are not restricted to muddy substrate but can also grow on coral or rocky substrate. The most unsuitable substrate is sand, which indicates high-energy water movement, in which new seedlings of mangroves cannot become fixed within one tidal cycle (necessary for development). Mangroves are also not restricted to salt water but can thrive in brackish and even fresh water, as can be seen in the Everglades of Florida, 30 km from the open sea. Mangroves are not restricted to belts along tropical and subtropical shorelines, but exhibit a wide range of habitats (see Fig. 6.18–6.32), mostly depending on tidal cycles and climate. The exposed outer fringes of a mangrove stand may show other species than the inner forest parts, which are protected from waves. Extreme landward areas with limited seawater intrusion and high evaporation rates contain specialized species that exhibit shrub-like growth, or mangroves may be absent due to salinity levels too high to sustain survival; these habitats are the so-called "tans" or "sabkhas" (Figs. 6.21–6.23, and 6.28–6.32).

Mangroves, which provide very important ecosystems for many organisms (crabs and fishes, birds, insects), have been eradicated over recent decades to transform low lying coasts into land for aqua-farming, which forms the base of profitable crab, oyster, lobster or fish industries. Dense natural mangrove forests, however, are also protective belts against storm waves and abrasion, but their potential as protection against tsunamis is under debate. Several countries with formerly dense mangrove forests have started to replant samplings and disperse seeds to stabilize coastlines.

6.18

6.19

Fig. 6.20 In the Northern Territory of Australia near Darwin many different mangrove landscapes are developed. This example, photographed during high water, clearly shows a differentiation of species from the outer rim to inland. Oblique aerial photograph (Photo credit: S. Scheffers).

Fig. 6.21 Where water is in short supply, mangroves may form gallery rims along tidal creeks as in the Northern Territory of Australia. Oblique aerial photograph, (Photo credit: S. Scheffers).

◀ Fig. 6.18 Dense mangrove forests on the Drowned Cays island west of Belize Town on the Caribbean coastline of Honduras, Central America at about *17° 28′N* and *88° 05′W*. Scene is about 1.5 km wide. (Photo credit: ©Google Earth 2010).

◀ Fig. 6.19 At the south coast of Puerto Rico (Caribbean) the growth and differentiation of mangrove species is very well developed, in particular in lagoons open to the sea (*17° 56′N* and *66° 14′W*, width of scene is 1.2 km). (Photo credit: ©Google Earth 2010).

Data SIO, NOAA, U.S. Navy, NGA, GEBCO
©2010 Google

6.22

Image © 2010 DigitalGlobe
©2010 Google

6.23

Fig. 6.24 Mangroves do not only settle in silt and clay, but also on hard substratum (e.g. a fringing coral reef), as this pictures shows from the Darwin area, Northern Territory, Australia. (Photo credit: S. Scheffers).

Fig. 6.25 A "mangrove parkland" with open space along the Queensland coast of Australia. The rough patches in the sand are pills formed by the crab *Dotilla*. (Photo credit: S. Scheffers).

◀Fig. 6.22 Mangrove landscape with wide bare areas as hypersaline tidal flats (tanns). In their greyish parts seaward algae and bacteria cover them, but in the very white parts life is nearly absent. (NW Australia) at *17° 15′S* and *123° 41′E*. Width of scene is 22 km. (Photo credit: ©Google Earth 2010).

◀Fig. 6.23 Another pattern of mangrove distribution with catenas to bare tanns of NW Australia. Location is *15° 32′S* and *128° 07′E* along a river length of nearly 20 km inland. (Photo credit: ©Google Earth 2010).

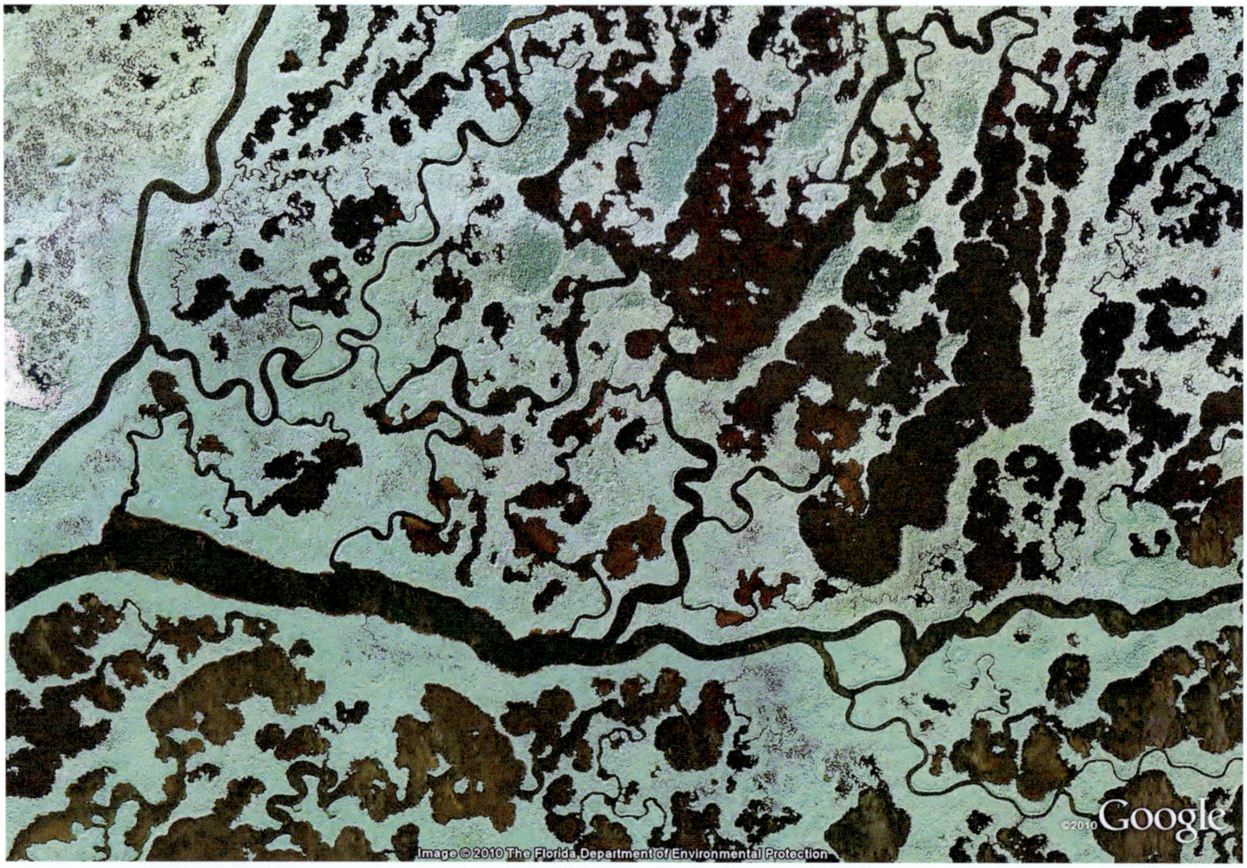
Image © 2010 The Florida Department of Environmental Protection
©2010 Google

6.26

Image © 2010 TerraMetrics
Image NASA
©2010 Google

6.27

Fig. 6.28 Many mangrove landscapes show a high differentiation of growth forms, species distribution and bare hypersaline tidal flats ("tanns") as in this example from the Darwin area of the Northern Territory, Australia, photographed from a plane during high water. (Photo credit: S. Scheffers).

These images and the satellite views show the great variety of mangrove landscapes in tropical environments with seasonal rainfall. Most images are from the Northern Territory (Australia) and the northernmost part of Western Australia. Mangroves are concentrated along coastlines and tidal creeks, but also colonize riverbanks, where they grow on the sheltered sides of river bends. The colour of the bare hyper-saline tidal flats depends on the salt content: the higher the salt concentration, the brighter and whiter the colors of the tanns. In waters with low salt concentrations a light grey cover of algae can be present. Important to note is that most areas presented here are in a macro-tidal environment with a five or more meters spring tide range. The highest part of the intertidal zone with the highest salt content is at the height of the two equinoctial high floods in spring and autumn.

◄Fig. 6.26 The "Ten Thousand Islands" in western Florida (USA) is a labyrinth of mangrove patches and tidal creeks at *25° 51′N* and *81°28′W*. Width of image is about 5 km. (Photo credit: ©Google Earth 2010).

◄Fig. 6.27 Extended mangrove forests with numerous tidal creeks in the coastal plain of southern Papua-New Guinea at *5°07′S* and *137°36′E*. Width of image is 40km. (Photo credit: ©Google Earth 2010).

NEXT PAGE

►Fig. 6.29 The width of this amphibious mangrove-and-tann zone is nearly 30 km from the open sea to the terrestrial environment. This region is in the Northern Territory of Australia at about *15°S* and *129°E* with a width of the scene of about 40km. (Photo credit: ©Google Earth 2010).

►Fig. 6.30 Northern Territory of Australia with an amphibious mangrove landscape at about *12°12′S* and *136°21′E*. Width is 10 km. (Photo credit: ©Google Earth 2010).

►Fig. 6.32 Along tidal creeks, which have connection to fresh water rivers, mangroves grow far inland along the river banks (west coast of the Gulf of Carpentaria, Australia, at about *15°51′S* and *136°39′E*, 3 to 25 km inland). (Photo credit: ©Google Earth 2010).

Image © 2010 DigitalGlobe
Image © 2010 TerraMetrics
©2010 Google

6.29

Image © 2010 DigitalGlobe
Image © 2010 TerraMetrics
©2010 Google

6.30

Mangrove species distribution

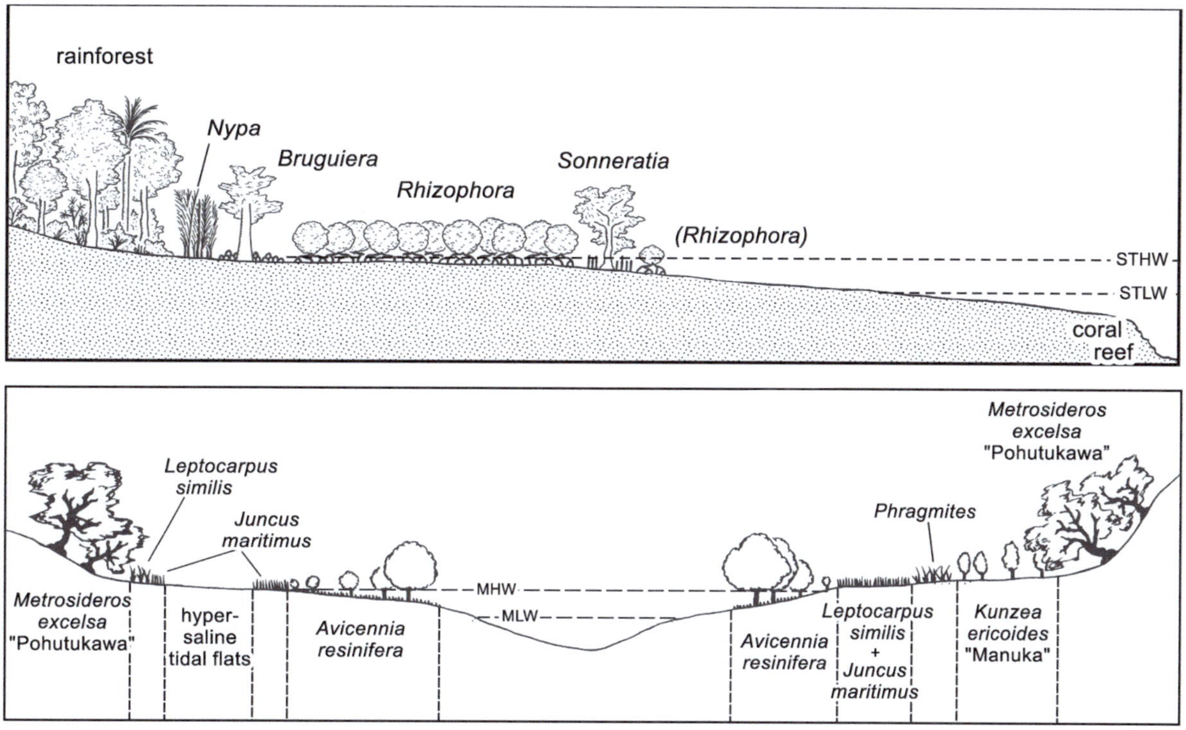

Fig. 6.31 a,b Two catenas of mangrove species distribution, a) Palau, Micronesia, in the centre of mangrove habitats. b) the southern fringe of mangrove habitats on the North Island of New Zealand (Kelletat, 1987; 1991).

6.32

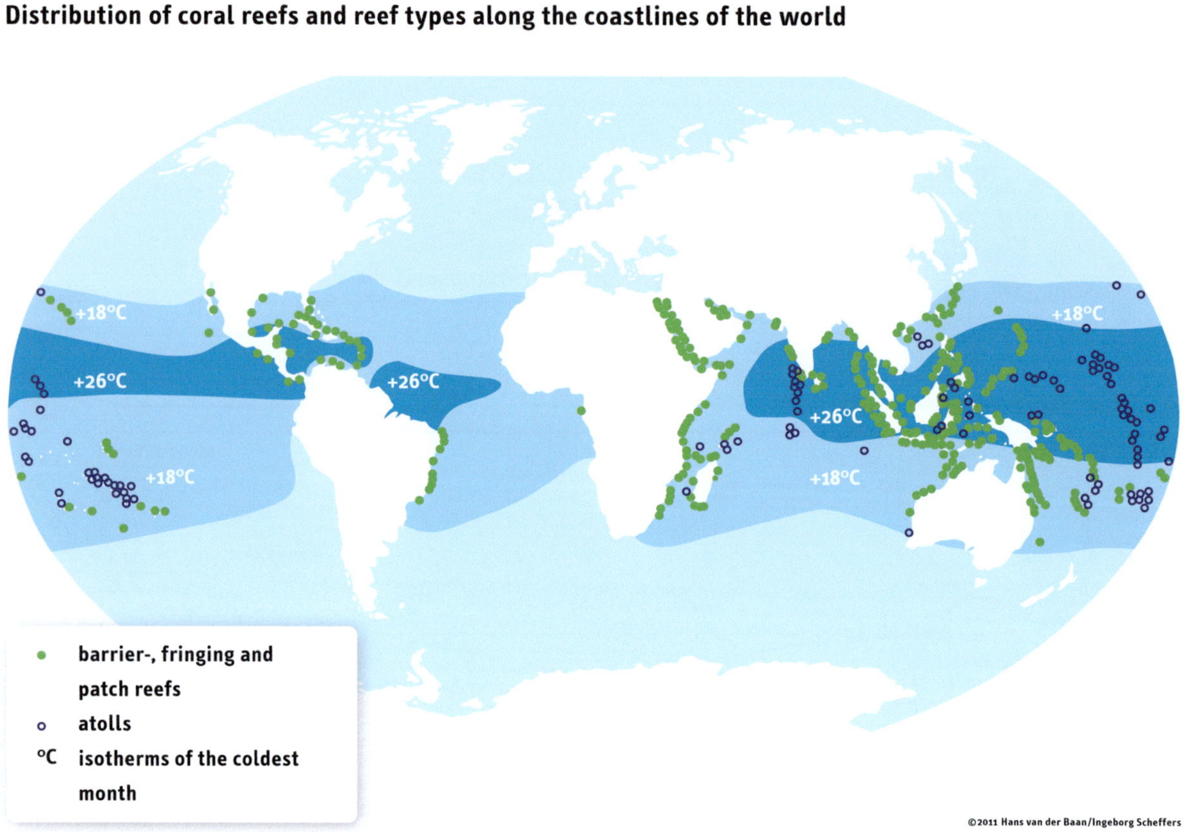

Fig. 6.33 Distribution of coral reefs and reef types along the coastlines of the world, and the isotherms of 18°C and 26°C for seawater in the coldest month of the year. (Image credit: Unesco)

6.3 Coral Reefs

Besides mangrove stands and sea grass areas, there are a number of other coastal environments that are dominated by organisms. We have already mentioned destructive systems (bio-erosive systems), but much more evident are those systems where constructions of hard limestone material, resistant to wave action are formed. Some are banks of mussels or calcareous algae, others are benches formed by vermetids and barnacles, but the most prominent systems are the coral reefs (Short & Woodroffe, 2009; Spalding et al., 2001; Veron 2000; see also Figs. 6.33 and 6.34).

Coral reefs are bio-constructions of hard limestone (Figs. 6.35–6.44). The most important reef builders are the stony (hermatypic) or hard corals (Order: Scleractinia, Class: Anthozoa) and the skeletons of these tiny coral polyps form the reefs. Reef-building corals have evolved a crucial symbiotic relationship with a genus of brown algae called zooxanthellae (Symbiodinium sp.). Millions of these single-celled algae are living as symbionts within the coral polyp's tissues. Zooxanthellae produce sugars and oxygen through photosynthesis thus helping the coral (amongst others) in the process of producing limestone or calcium carbonate. Corals with zooxanthellae grow up to three times faster than corals without zooxanthellae. A coral polyp consists of a fleshy sack with tentacles around the opening, or mouth, which sits in a limestone (aragonite) skeletal cup, secreted by the polyp. Pores between polyps are gradually filled with small grains of abraded carbonates. All this material is cemented and is strong enough to withstand waves.

Coral reef growth depends on a number of environmental factors, and the two most important factors include sunlight (for symbiont photosynthesis), and enough space to expand. What this means is that coral reefs grow very fast in some periods, while in other periods they "turn their

Warm and cold water coral reefs

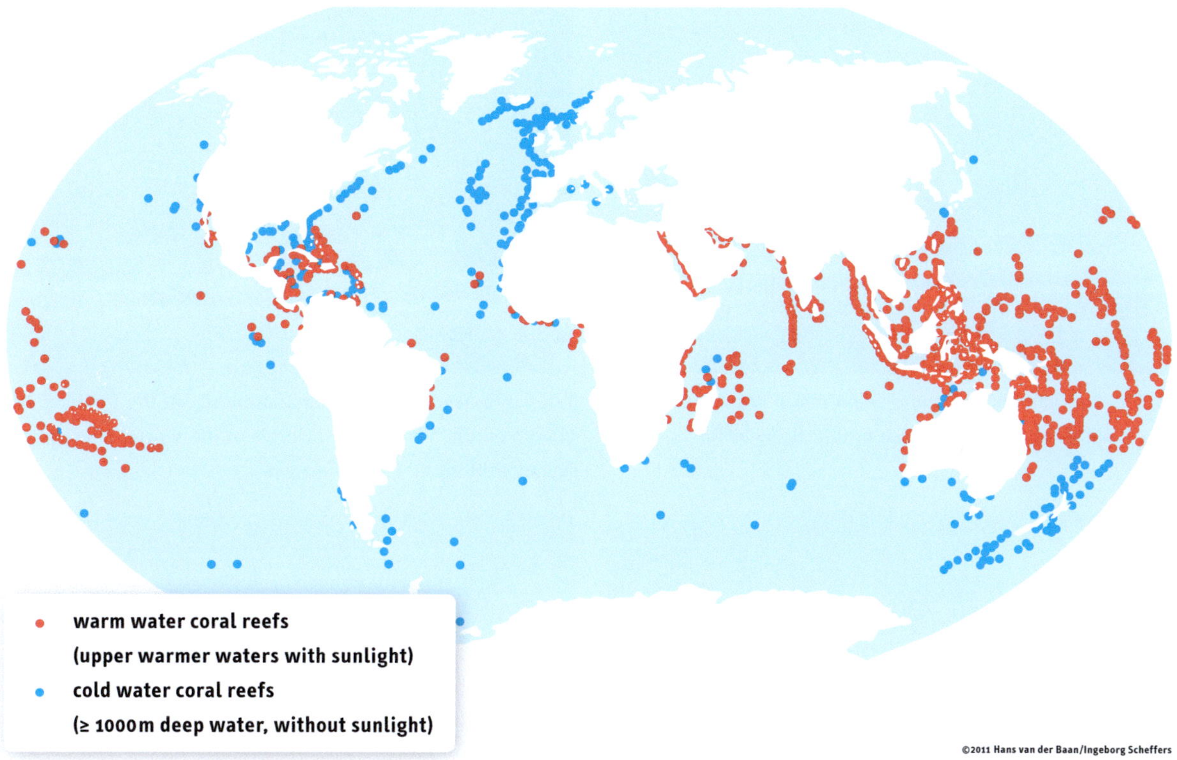

Fig. 6.34 In contrast to the warm water coral reefs, which are confined to the upper warmer waters due to the requirement for sunlight, the cold-water corals grow in depths of 1000 m and more without any light. Their existence was only discovered in recent years and distribution maps are scarce. (Image credit: Unesco)

growth off". Coral reefs can grow extensively in clear water without much suspended matter where sunlight penetrates deep into the water column. Therefore, coral reefs or reef communities close to coasts with extensive run-off (wet-tropics and areas with river plumes passing) will develop much slower or are even growth inhibited compared to reefs further from shore or in dryer climates. The growth rate of corals of several cm/year is normally high enough to catch up with changing (rising) sea level, and under optimum conditions growth seaward and against the surf occurs, resulting in reef flats of many hundreds of meters wide.

Fig. 6.35 A close up image of the Caribbean brain coral *Diploria strigosa* in Curaçao (Photo credit: S. Scheffers).

Fig. 6.36 Caribbean brain corals *(Diploria strigosa)* (Photo credit: S. Scheffers).

Fig. 6.37a A Caribbean *Acropora palmata* branching coral (Elkhorn coral) at Curaçao (Photo credit: S. Scheffers).

Fig. 6.37b *Acropora palmata* community at the Cayman Islands (Photo credit: Veron J.E.N. Corals of the World, photography by Nancy Sefton).

Fig. 6.38 The delicate growth forms within this *Acropora sp.* dominated coral community points to a more sheltered environment. (Credit: Veron J.E.N. *Corals of the World,* photography by Robert Steene).

Fig. 6.39 These large *Porites lutea* corals in Southern Thailand are more than 500 years old. (Photo credit: S. Scheffers)

There are hundreds of different species of stony coral (Veron, 2000), especially in the "hot spot" area of SE Asia; the so called "coral triangle". This area, enclosed by Malaysia, Philippines, Solomon Islands and Indonesia is where scleractinian biodiversity is highest (over 500 species or 75% of global stony coral species). However, on a reef other bio-constructing organisms such as calcareous coralline algae, vermetids, bryozoans, foraminifers, and serpulids also contribute to reef construction. A reef's final form is the result of bioconstruction and the always-ongoing bio- and mechanical erosion. Flourishing and healthy reefs regenerate relatively fast after strong cyclonic impacts, as demonstrated by their survival in the Great Barrier Reef of Australia, in the South China Sea and around the Philippines. More than 2000 kilometers in length, the Great Barrier Reef on the east coast of Australia is considered the largest form ever built on Earth by organisms, yet new deep sea research has found much longer cold water reefs (but not as massive and extensive) in depths around 1000 meters exist from Norway to southern Africa. This is a much larger range and extent than the Great Barrier Reef. These reefs consist of very slow growing, calcium carbonate forming corals that lack zooxanthellae and are entirely dependent on organic material (from the surface and zooplankton) for their energy supply. There are six known cold (or deep) water coral species of which Lophelia pertusa is the most commonly found.

Coral reefs in the photic zone (i.e. shallow water zone within reach of sunlight) need special environmental conditions for optimal growth. One of these is warm water, never below 18°C (but not above 31°C), another is normal salinity (25% to 42%), clear water conditions (i.e. limited suspension load from river mouths), and enough light (i.e. reef corals can only flourish in shallow water up to depths of 30 m in large numbers, up to 60 m in smaller diversity, and even in very deep, dark and even cold water only as single corals (but here they do not form reefs). However, recent discoveries of tropical reefs and coral communities in highly turbid and estuarine waters show that corals can thrive under a wider variety of environmental factors than previously thought.

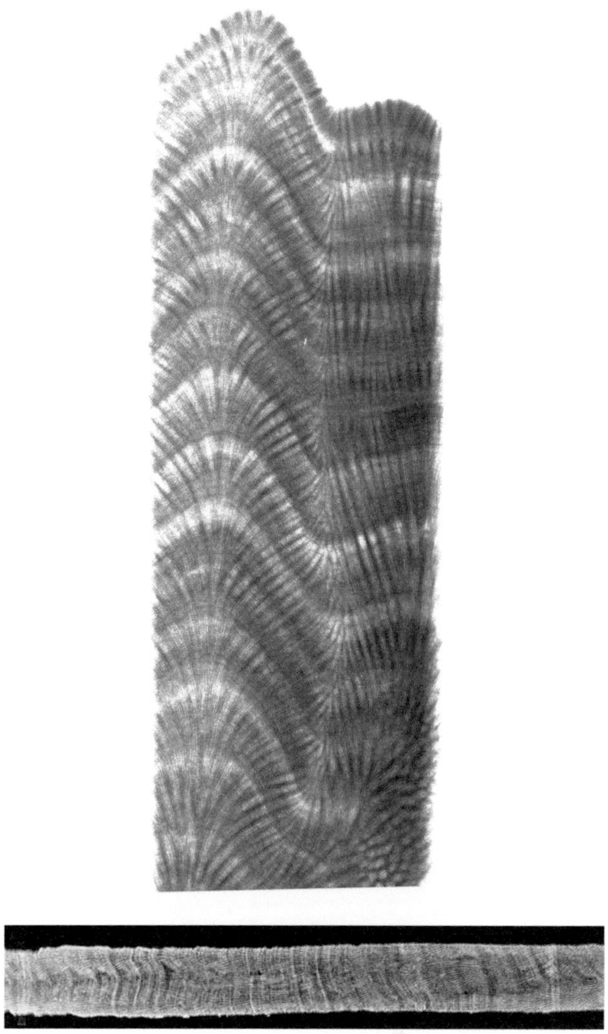

Fig. 6.40 Growth bands in corals allow reconstruction of environmental data and processes such as climate variability and sea surface temperatures. Therefore they provide historical perspectives on human impacts on coral reefs. (Photo credit: Thomas Felis, University of Bremen)

Fig. 6.41 A highly diverse Coral Reef in Micronesia. (Photo credit: Suchana Chavanich, Chulalongkorn University/Marine Photobank).

Fig. 6.42 Species-rich coral fringing reef off the Australian east coast exposed during extreme low tide. (Credit: Veron J.E.N. Corals of the World)

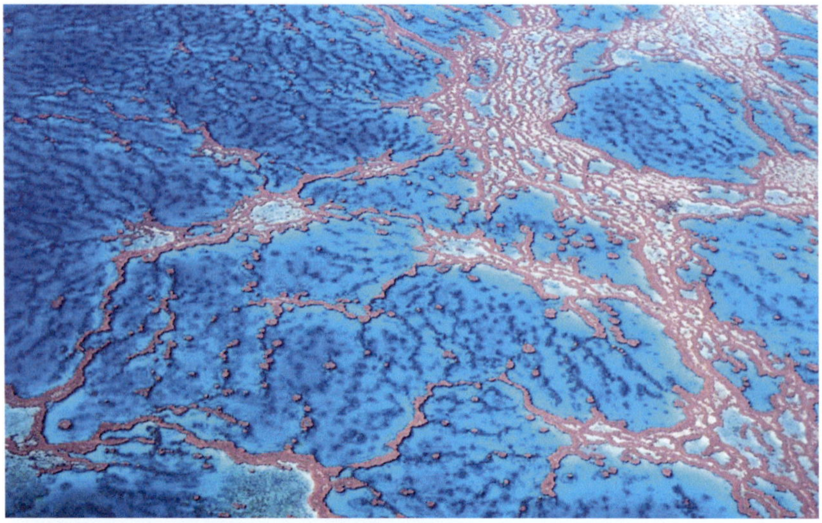

Fig. 6.43 Aerial photography of a reef flat on the Australian Great Barrier Reef during low water. (Photo credit: S. Scheffers).

▶Fig. 6.45 The asymmetrical Hook and Hardy Reefs of the Great Barrier Reef, growing outwards against the surf that is created by eastern trade winds, show a secondary growth pattern in their lagoons. 30 km wide scene at *19°47′S* and *149° 12′E*. (Photo credit: ©Google Earth 2010).

▶Fig. 6.46 Sometimes fringing reefs grow in coastal inlets as in this "sherm" off Egypt (Red Sea coast). Width of picture is 7 km. (Photo credit: ©Google Earth 2010).

Fig. 6.44 Aerial photography of a reef flat at Hardy Reef, Great Barrier Reef, Australia, during low water. (Photo credit: S. Scheffers).

Coral reefs are divided into four main types: fringing reef, platform reefs, barrier reefs and atolls (Figs. 6.45–6.72, see also Spalding et al., 2001). Fringing reefs are relatively young. They develop in shallow waters parallel to the coast of tropical islands or continents. The corals grow upwards to just below sealevel and outwards towards the open ocean. Fringing reefs are generally situated a short distance from shore and do not display a significant lagoon. Platform reefs are usually situated in relatively sheltered water and rather far offshore. These reef types exhibit horizontal crest with small and relatively shallow lagoons. A coral reef growing parallel to the coastline and separated from it by a lagoon is called a barrier reef. As the reef continues to grow further offshore it sooner or later reaches the edge of the continental shelf. Barrier reefs can originate offshore if the

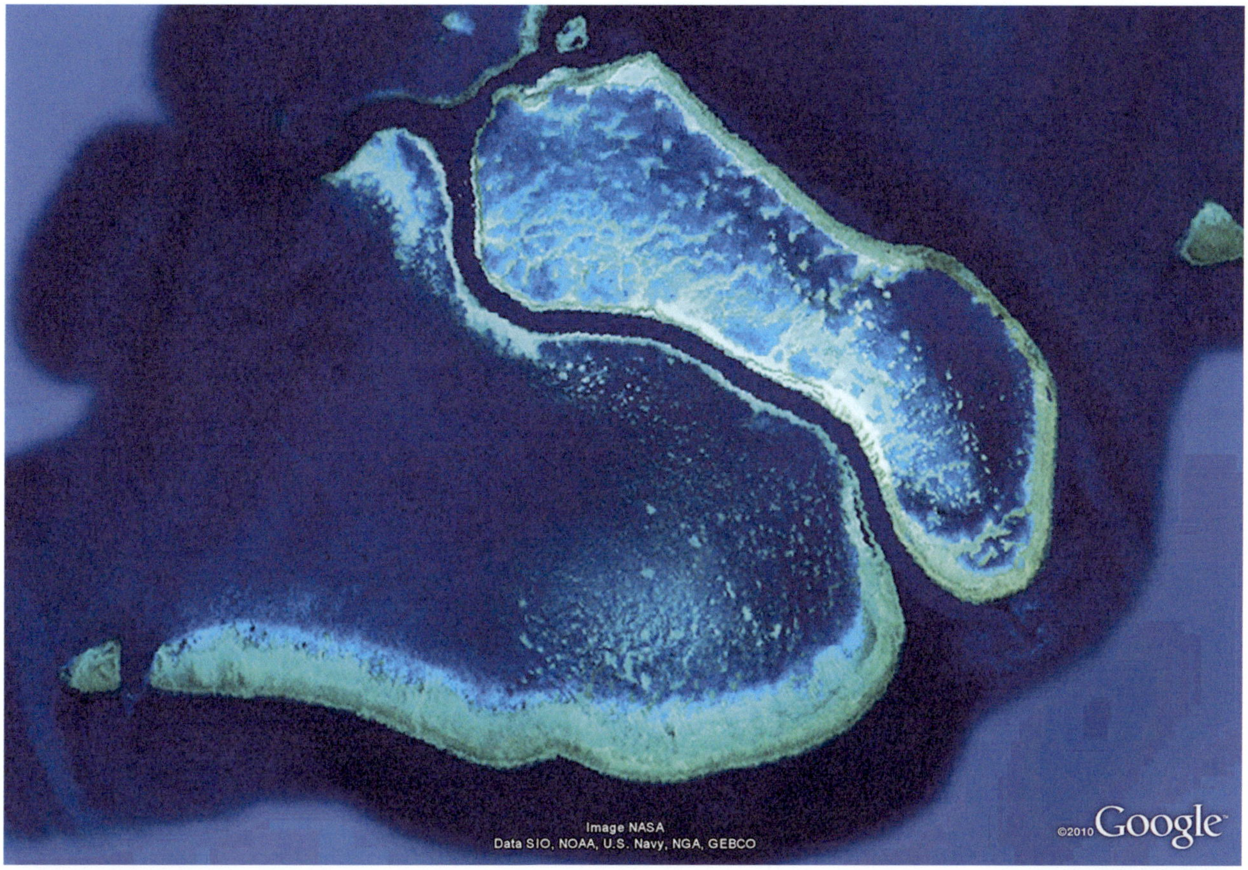
Image NASA
Data SIO, NOAA, U.S. Navy, NGA, GEBCO
©2010 Google

6.45

Image © 2010 DigitalGlobe
©2010 Google

6.46

Image © 2010 TerraMetrics
Image © 2010 GeoEye
Data SIO, NOAA, U.S. Navy, NGA, GEBCO
©2010 Google

6.47

Image © 2010 DigitalGlobe
Data SIO, NOAA, U.S. Navy, NGA, GEBCO
©2010 Google

6.48

Fig. 6.49 The fringing reefs of the Palau archipelago in Micronesia show sink holes and old channels pointing to a pre-Holocene base overgrown in the Younger Holocene. (Photo credit: S. Scheffers).

◀Fig. 6.47 A typical fringing reef along the north coast of Cuba at *20°44′N* and *75°26′W*. Width of image is 14 km. (Photo credit: ©Google Earth 2010).

◀Fig. 6.48 Fringing reef at the south coast of New Caledonia at *21°34′S* and *165°20′E*. Width of scene is 30 km. (Photo credit: ©Google Earth 2010).

Fig. 6.50 Aspects of a complex growth pattern in a fringing reef at Palau, Micronesia, indicating an older karst relief with sinkholes. (Photo credit: S. Scheffers).

depth of the seabed present is shallow enough at some stage during sea level rise to allow for corals to grow. However, the most famous barrier reef, the Great Barrier Reef, is hardly a barrier reef but a collection of platform reefs which developed on drowned topography paralleling the present-day coastline. Atolls (Figs. 6.68–6.73) are rings of reef, often surrounding an island (or a former volcano which is now submerged due to sea level rise, or has eroded or sunk, with the persistence of coral growth). Atolls typically have a shallow, sandy, protected lagoon in the middle. Direct contact to the open sea is through channels in the reef structure. The formation of atolls was under debate in former times, but the hypothesis of Charles Darwin more than 150 years ago is now widely accepted: Darwin concluded that the circular nature of an atoll points to a round island (e.g. of volcanic ori-

Fig. 6.51 *Tridacna sp.* (Giant clams) from Southern Thailand is abundant on patch- and fringing reefs. Their size may reach more than 1 m across. The colours originate from pigments of symbiotic algae inhabiting the tissue of the clams. (Photo credit: S. Scheffers).

Fig. 6.52 An atoll in the northern part of the Palau islands, Micronesia. A gap in the island chain allows storms to transport huge amount of coral sand into the central lagoon. (Photo credit: S. Scheffers).

gin), where the reef first grew as a fringing structure. By subsidence of the volcanic edifice, due to isostatic response to the weight of the lava (and also rise in sea level), the reef growth kept pace with the (relatively) rising sea level, whereas the island slowly disappeared. What remained after the volcano disappeared from view was a ring of coral reefs surrounded by a deep ocean. Other theories suggest that corals merely colonized isolated hills that were flooded by rising sea level.

Another morphological feature of coral reefs are the clearly defined zones:

- A reef flat a shallow, level section uncovered at low tide with local protected deeps in which living corals flourish, often partly damaged by storms. They can range in area up to some square kilometers. Depending on their history of development, wave action and sea surface temperatures, the surface may be featureless or a complex maze of crevices.
- A reef front or reef crest where the reef takes the full force of wave action. Growth of corals is generally restricted by storm damage and usually there is a lot of rubble and sand, with a high quantity of calcareous algae.
- An outer reef slope the seaward edge of a reef is sometimes steep and slopes down to deeper water. Since the water is generally clearer, abundant coral growth may be present to depths of 50 m depending on available light. Often present are caves, overhangs or gullies. However, some

Fig. 6.53 Coral reefs grow against the steady surf produced by trade winds. The waves provide an oxygen rich environment and a steady nutrient supply (Palau, Micronesia). (Photo credit: S. Scheffers).

Fig. 6.54 a,b,c,d A sequence of reef evolution: From fringing reefs towards barrier reefs. Lagoon width and distance from the mainland increase due to island subsidence. From Moorea and Bora Bora, French Polynesia. (Photo credit: S. Scheffers).

Image © 2010 TerraMetrics
Data SIO, NOAA, U.S. Navy, NGA, GEBCO
Image © 2010 DigitalGlobe
©2010 Google

6.55a

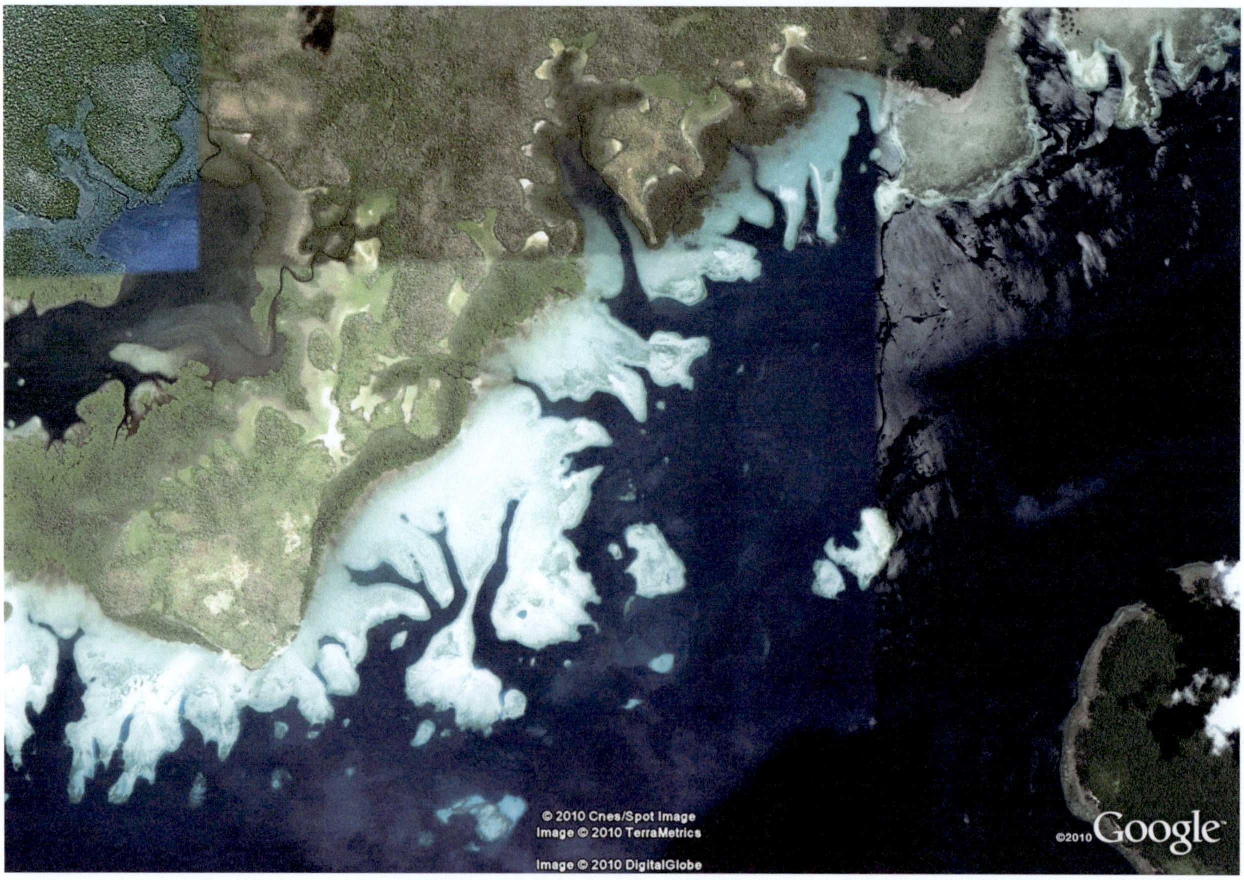
© 2010 Cnes/Spot Image
Image © 2010 TerraMetrics
Image © 2010 DigitalGlobe
©2010 Google

6.55b

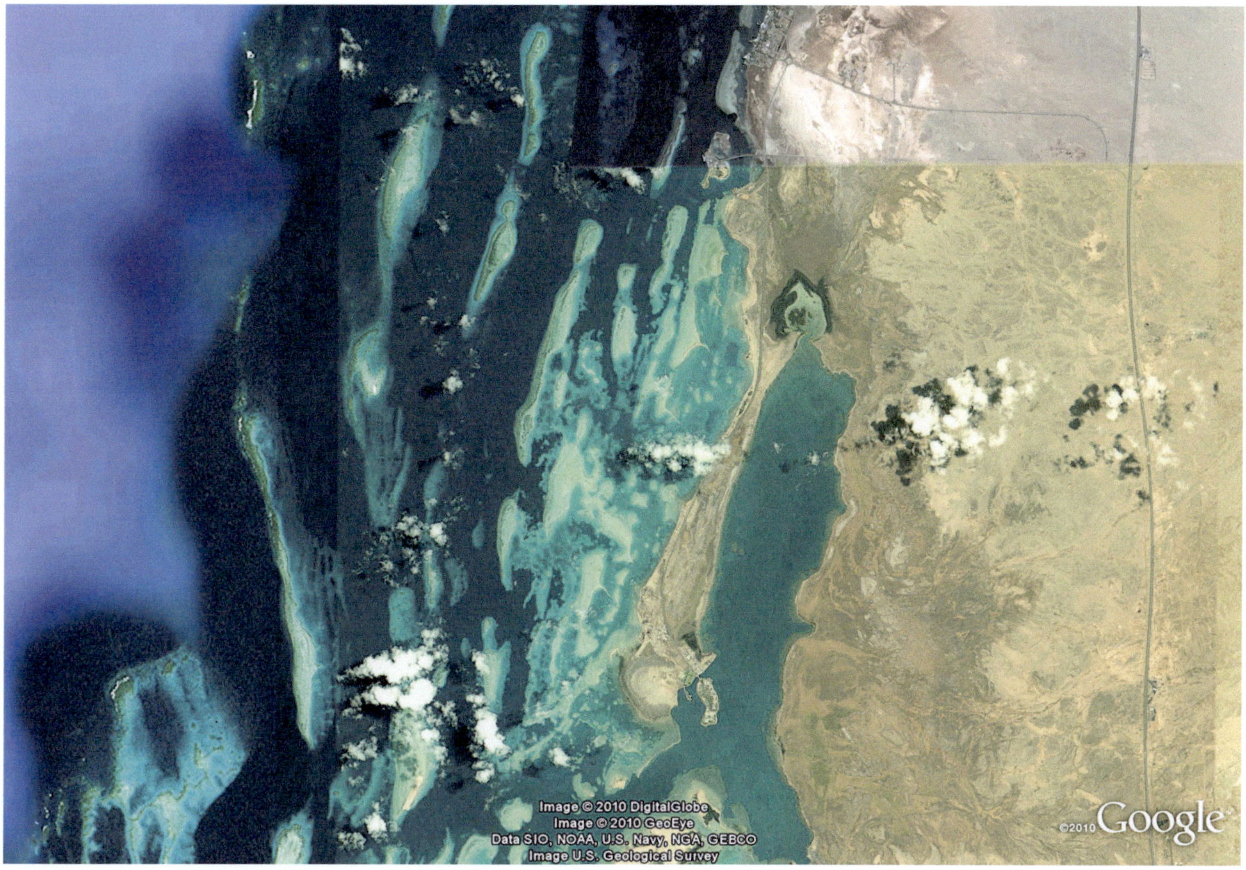

6.56

Fig. 6.56 These parallel reefs grow on parallel submerged and cemented dune ridges (Red Sea, coast of Saudi Arabia at *23° 14′N* and *38° 42′E*, scene 10 km wide). (Photo credit: ©Google Earth 2010).

outer reef slopes are nearly vertical, the so-called drop-offs or walls. Sand and sediment will only collect in terraces or shelves. There might be vertical chutes with sand and debris including collapsed blocks of the reef front.

The form of a coral colony points to the place where it lives. In general, more delicately branched and plated species are mostly from sheltered sites; those with thick branches or massive forms grow in more exposed places on the reef. The coral reefs of the world are under pressure from many natural and anthropogenic disturbances: natural processes are intense storms and tsunamis as well as unusual fresh water discharges and associated suspended sediment run-off with contaminants from the mainland. Another natural disturbance is that of large temperature variations below 18°C and above 31°C due to ENSO events (= El Niño Southern Oscillation, a quasi-periodic climate pattern that occurs across the tropical Pacific Ocean approximately every five years). Coral bleaching, the expulsion of zooxanthellae due to extended periods of warmer than normal sea temperatures, has affected more than 50% of all coral reefs of the world in the last few decades, this may be a direct consequence of climate change.

◂ Fig. 6.55a,b Along several Indonesian islands (at *2° 15′N* and *118° 12′E* above, and at *14°N* and *122°E* below) we find these fringing reefs dissected by small valleys. This could point to a Pleistocene origin with a much lower sea level than at present. The modern living reef is just a thin layer on the older base. (Photo credit: ©Google Earth 2010).

Ocean acidification is not a symptom of climate change; rather, it is a threat concurrent with climate change and caused by a common root problem: ongoing anthropogenic CO_2 emissions. Acidification of the ocean waters affects many carbonate dependent organisms in a variety of ways (from decreased growth rates to inhibition of larval development).

Local human impacts are manifold, either directly or indirectly: direct ones are e.g. coral destruction by snorkelers and divers, collection of corals for the tourist market/aquarium trade,

Image © 2010 DigitalGlobe
Data SIO, NOAA, U.S. Navy, NGA, GEBCO
©2010 Google

6.57

Image © 2010 DigitalGlobe
© 2010 Cnes/Spot Image
Data SIO, NOAA, U.S. Navy, NGA, GEBCO
©2010 Google

6.58

Fig. 6.59 Elongated patch reefs in the Torres Strait north of Queensland, where strong currents meet between the Pacific and Indian Oceans. The region is at *10° 27′S* and *142° 11′E*, and the length of the reefs is up to 15 km. (Photo credit: ©Google Earth 2010).

dynamite and cyanide fishing and limestone burning. Indirect ones e.g. are eutrophication of the seawater from agricultural activities, or sediment run-off from land due to deforestation, just to name a few.

◄Fig. 6.57 Coral reefs along and close to the coastline are called fringing reefs, sometimes exhibiting deep lagoons in places of former depressions. Example from the Red Sea coast of southern Egypt at about *22° 16′N* and *36° 38′E,* scene is about 7 km wide. (Photo credit: ©Google Earth 2010).

◄Fig. 6.58 Patch reefs are the result of one coral colony growing upwards and outwards, subsequent die-off and replacement and/or resettlement of other coral larvae on the remaining skeleton, surrounded by a "halo" of sand. Red Sea of Egypt at the boundary to Sudan, *23° 11′N* and *35° 39′E,* 19 km wide. (Photo credit: ©Google Earth 2010).

Fig. 6.60 The outer reef zone of western Palau (Micronesia) is a well developed closed barrier reef. Oblique aerial photograph, (Photo credit: S. Scheffers).

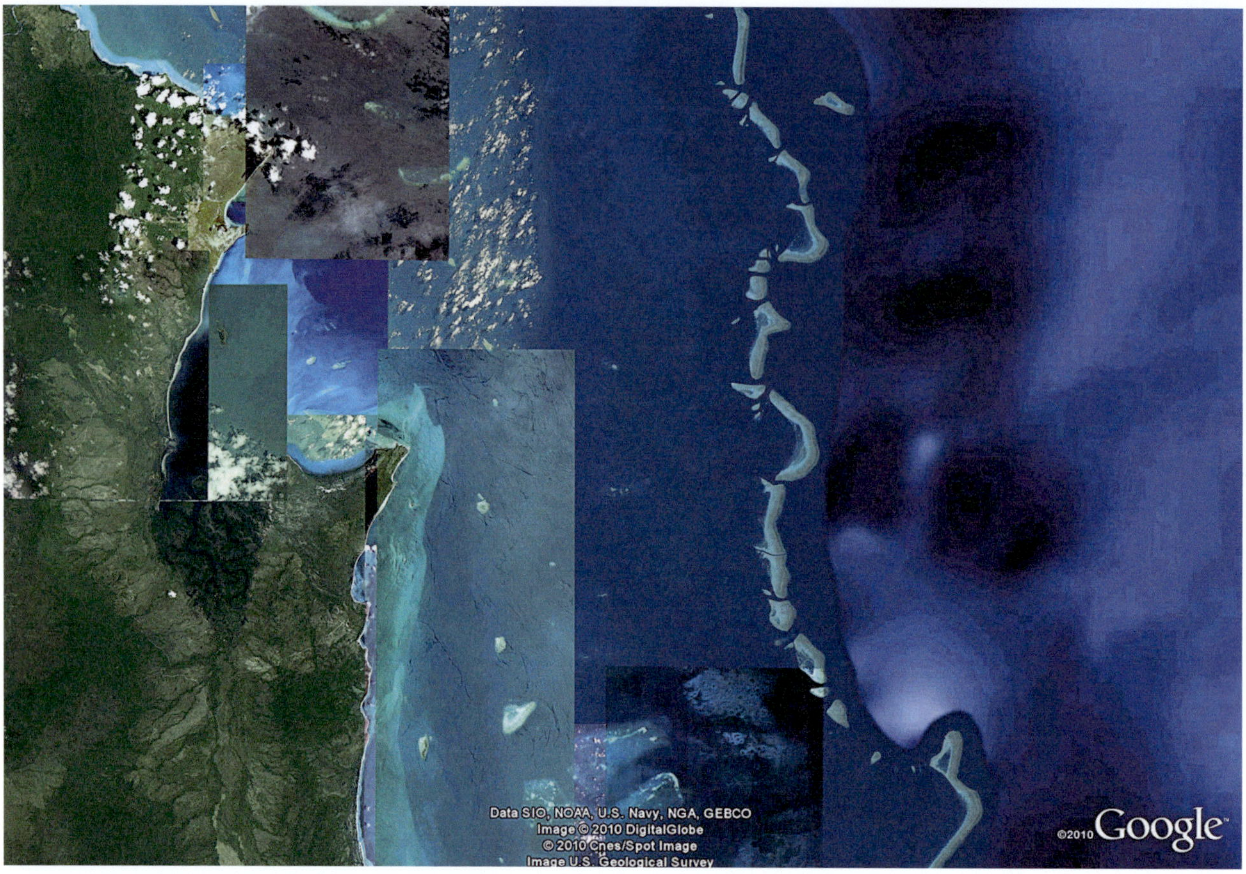

Fig. 6.61 Along the coastline of NE Australia the largest coral reef system on Earth, the Great Barrier Reef, is visible. This reef system has an average distance to the coastline of up to 50 km and is about 2000 km long. (Photo credit: ©Google Earth 2010).

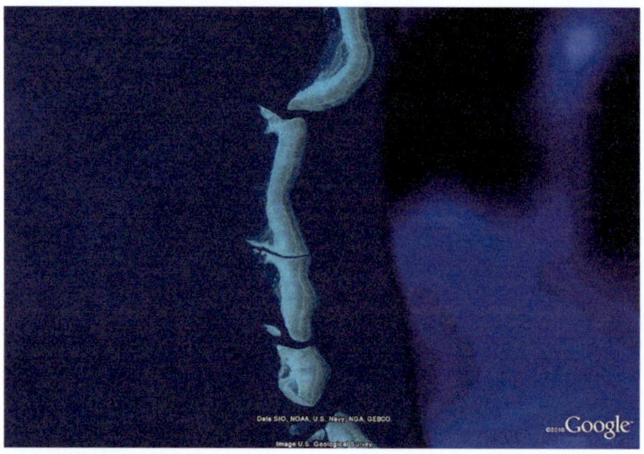

Fig. 6.62 Detail of the outer ribbons of the Great Barrier Reef. The section shown is about 15 km long, the ribbons up to 1 km wide, position is around *12° 53′S* and *143° 49′E*. (Photo credit: ©Google Earth 2010).

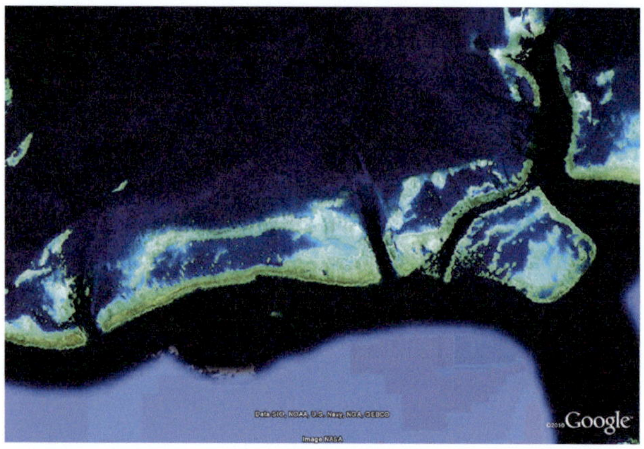

Fig. 6.63 A detail from the outer parts of the southern Great Barrier Reef shows several channels with deep water at *19° 13′S* and *148° 14′E*. Width of scene is 20km. (Photo credit: ©Google Earth 2010).

►Fig. 6.64 Parts of the Great Barrier Reef show complex growth patterns. The different water colours at low tide are an indication of water depth. Scene of 32 km wide at *21°S* and *151° 27′E*. (Photo credit: ©Google Earth 2010).

►Fig. 6.65 Another example of coral reef construction along the coast of eastern Australia at *21° 10′S* and *151° 38′E* with 60 km width. (Photo credit: ©Google Earth 2010).

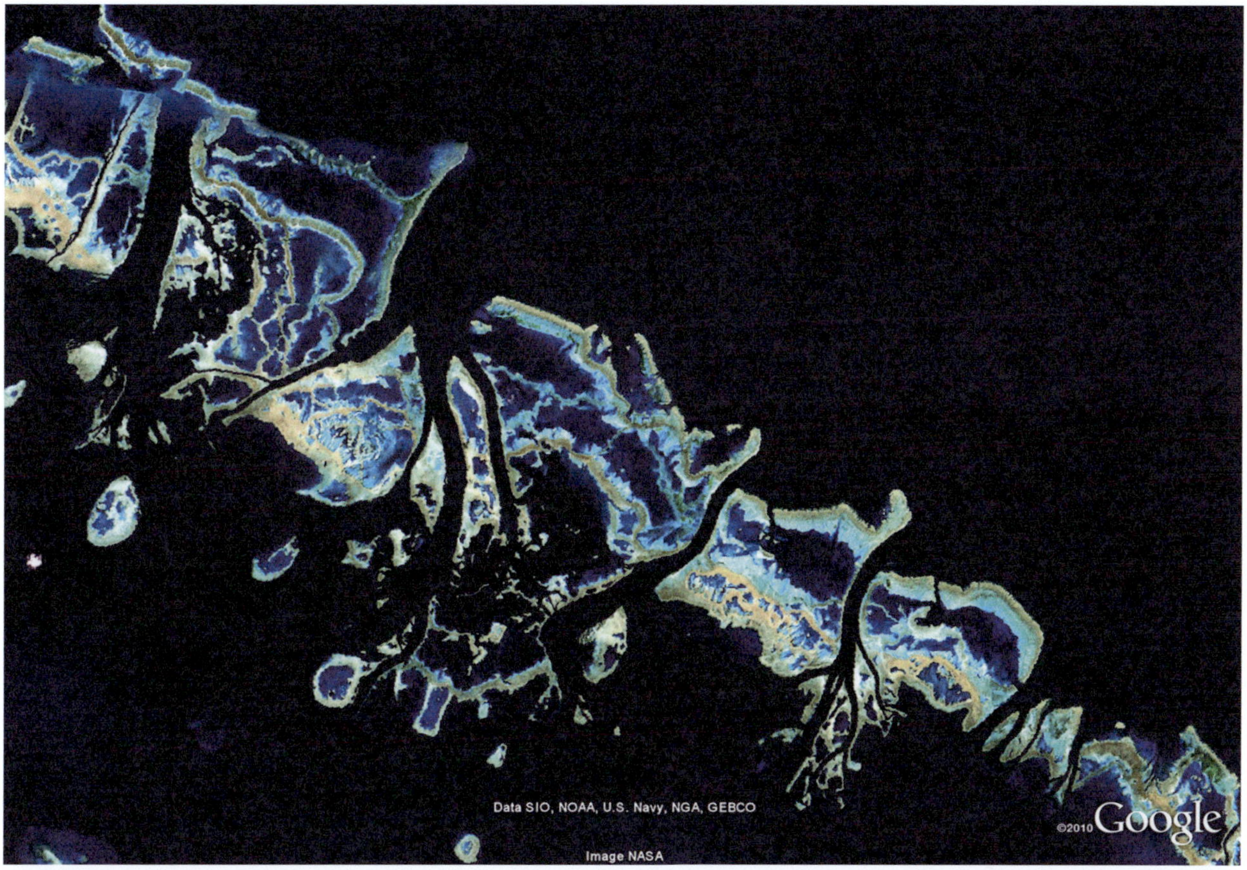
Data SIO, NOAA, U.S. Navy, NGA, GEBCO
©2010 Google
Image NASA

6.64

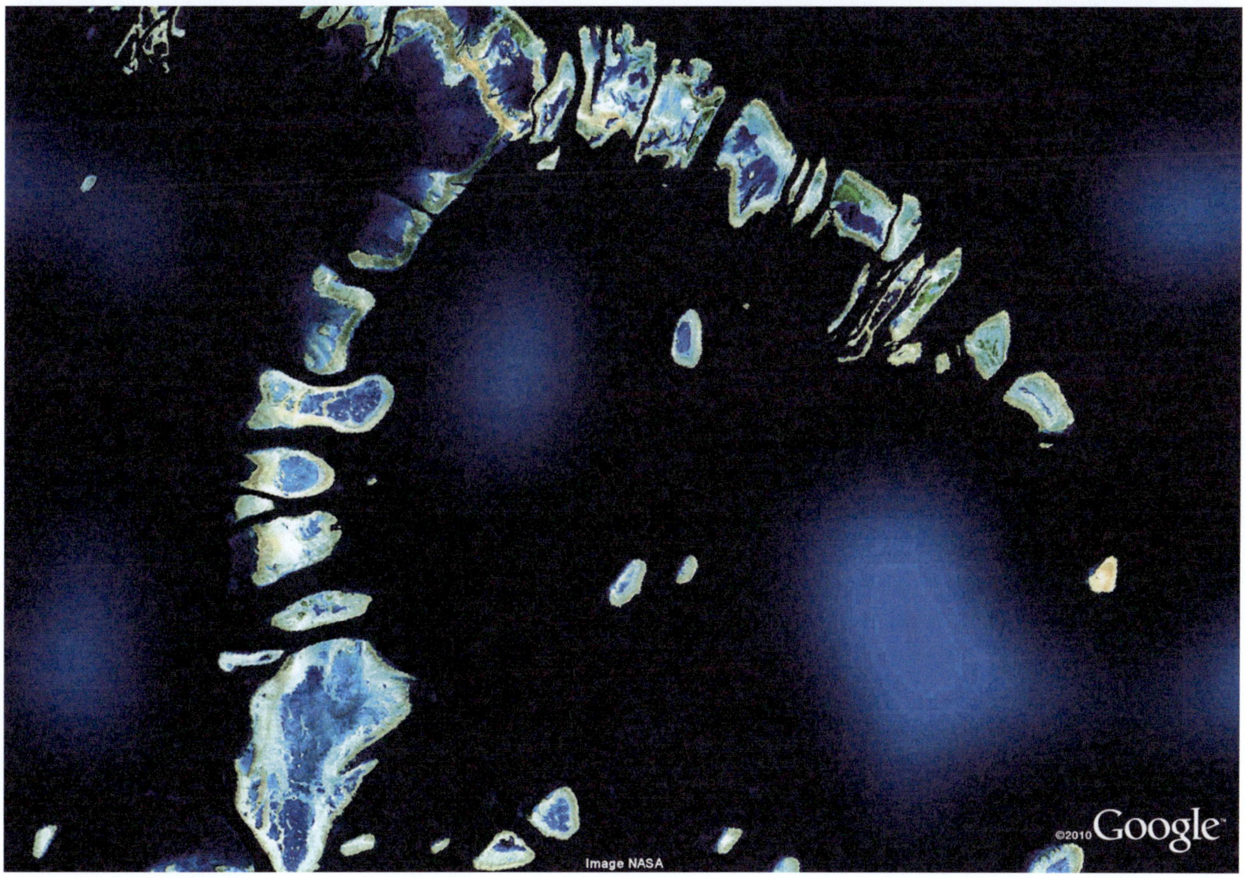
©2010 Google
Image NASA

6.65

Fig. 6.66 Moorea, north of Tahiti (French Polynesia) is an example of an inactive volcanic island surrounded by a fringing reef. This is the first stage in the development of an atoll in a hot spot island sequence. The diameter of this island including the fringing reef is a little less than 20 km at *17° 32′S* and *149° 50′W*. (Photo credit: ©Google Earth 2010).

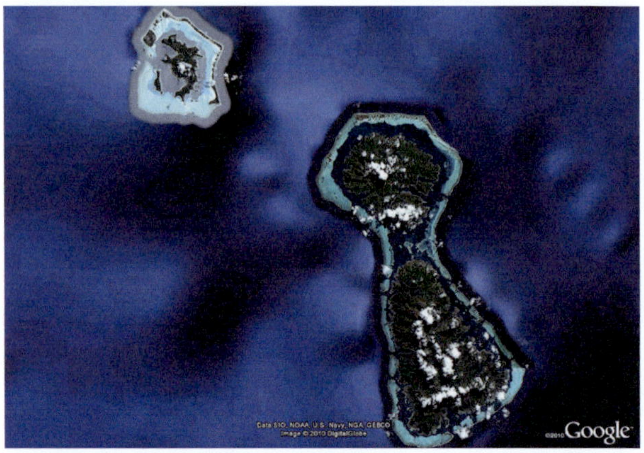

Fig. 6.67 Tahaa and Raiatea (north of Tahiti, more than 40 km long) show fringing reefs with clearly developed reef lagoons close to old volcanoes, pointing to an older stage of island subsidence, whereas Bora Bora in the north is in a state of a barrier reef development, and is assumed to be older than Tahaa and Raiatea. Location is around *16° 41′S* and *151° 33′W*. (Photo credit: ©Google Earth 2010).

Fig. 6.68 According to Darwin's hypothesis, a sinking and eroded volcanic island will lead to atoll formation. The image shows Mayotte island with a 40 km wide coral fringe in the western Indian Ocean at *12° 50′S* and *45° 06′E*. (Photo credit: ©Google Earth 2010).

Fig.6.69 Bora Bora, French Polynesia: an old volcanic edifice surrounded by a barrier reef. (Photo credit: Panoramio; http://www.panoramio.com/photo/9439).

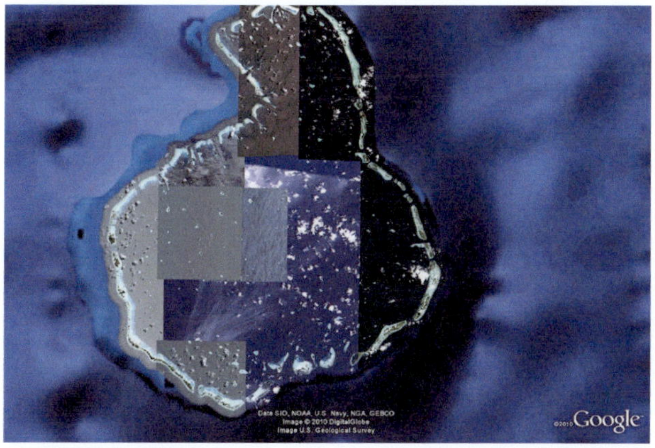

Fig. 6.70 A 65 km wide atoll in the southern section of the Maldives (Indian Ocean) at *0° 32′S* and *73° 18′E*. (Photo credit: ©Google Earth 2010).

►Fig. 6.71 a,b,c The Maldives atolls consist predominantly of smaller ring-like reefs, called "faros". Keadu is at about *3° 23′N* and *73° 32′E*, the scene is about 20 km wide. The lower images show enlargements of the outer asymmetrical faros, formed by dominant waves from the SW monsoons. The faros are oriented westwards. b) A 13 km wide section of the west frame of a Maldive atoll with faros, here at *2° 49′N* and *73° 22′E*. North is to the right, the island on the left faro is Kureli. c) A 6.5 km long island (Nilandu) of the Maldives (north is to the right), at *3° 18′N* and *72° 50′E*. (Photo credit: ©Google Earth 2010).

►Fig. 6.72 Ring-like coral reefs of all dimensions are typical for the Maldives in the northern Indian Ocean and are called "faros". They exhibit diameters of several hundred meters to several kilometers. Location is near Kureli at *2° 49′N* and *73° 22′E*. (Photo credit: ©Google Earth 2010).

Data SIO, NOAA, U.S. Navy, NGA, GEBCO
Image © 2010 DigitalGlobe
Image U.S. Geological Survey
©2010 Google

6.71a

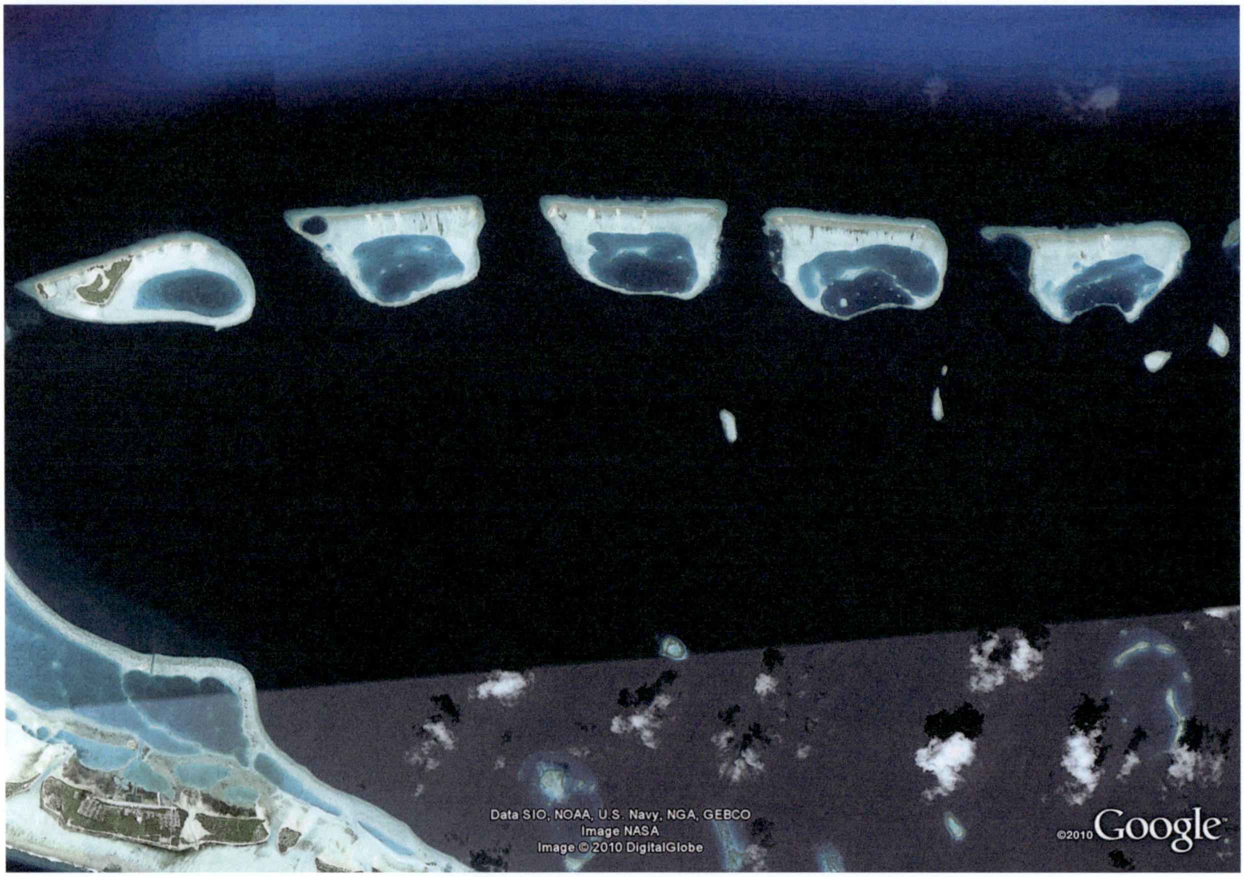
Data SIO, NOAA, U.S. Navy, NGA, GEBCO
Image NASA
Image © 2010 DigitalGlobe
©2010 Google

6.71b

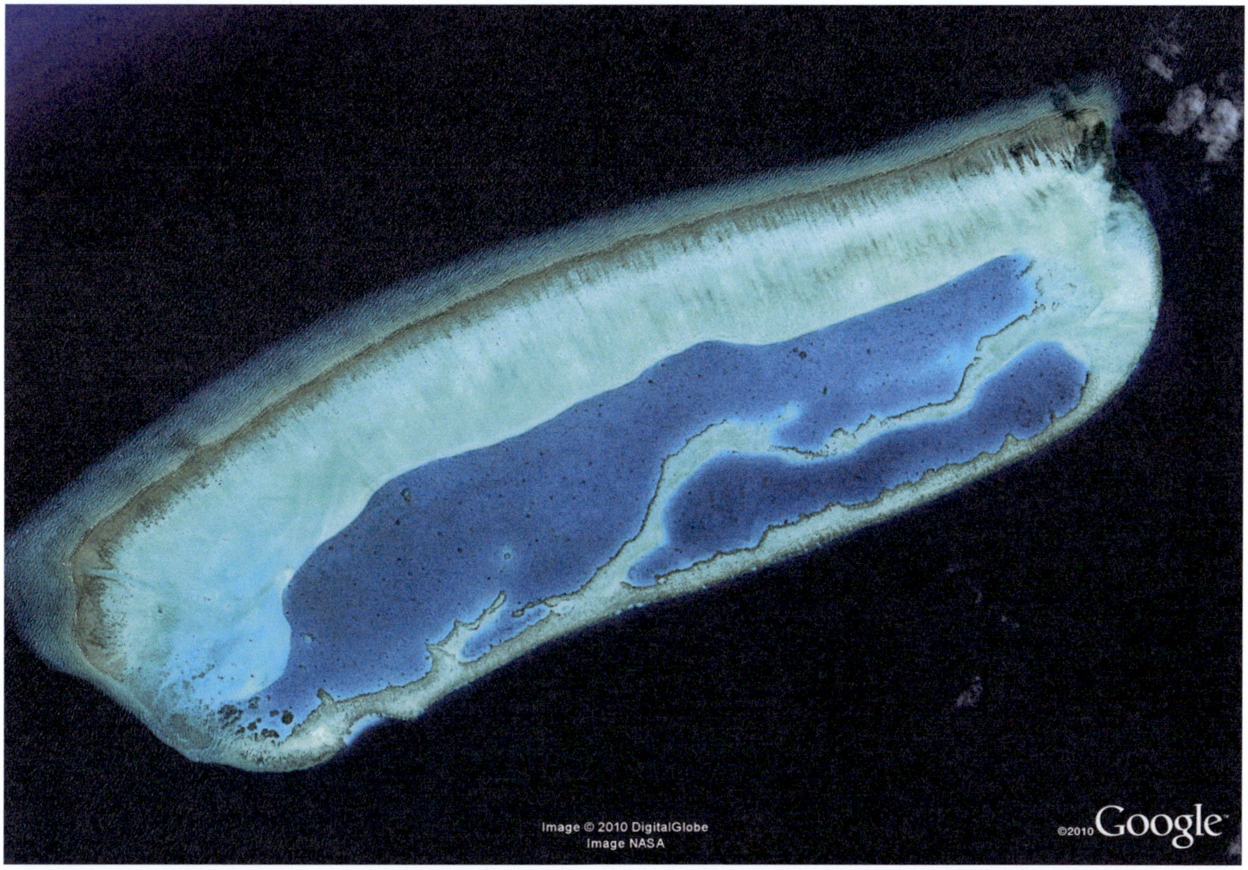
Image © 2010 DigitalGlobe
Image NASA
©2010 Google

6.71c

Image © 2010 DigitalGlobe
Image U.S. Geological Survey
Data SIO, NOAA, U.S. Navy, NGA, GEBCO
Image NASA
©2010 Google

6.72

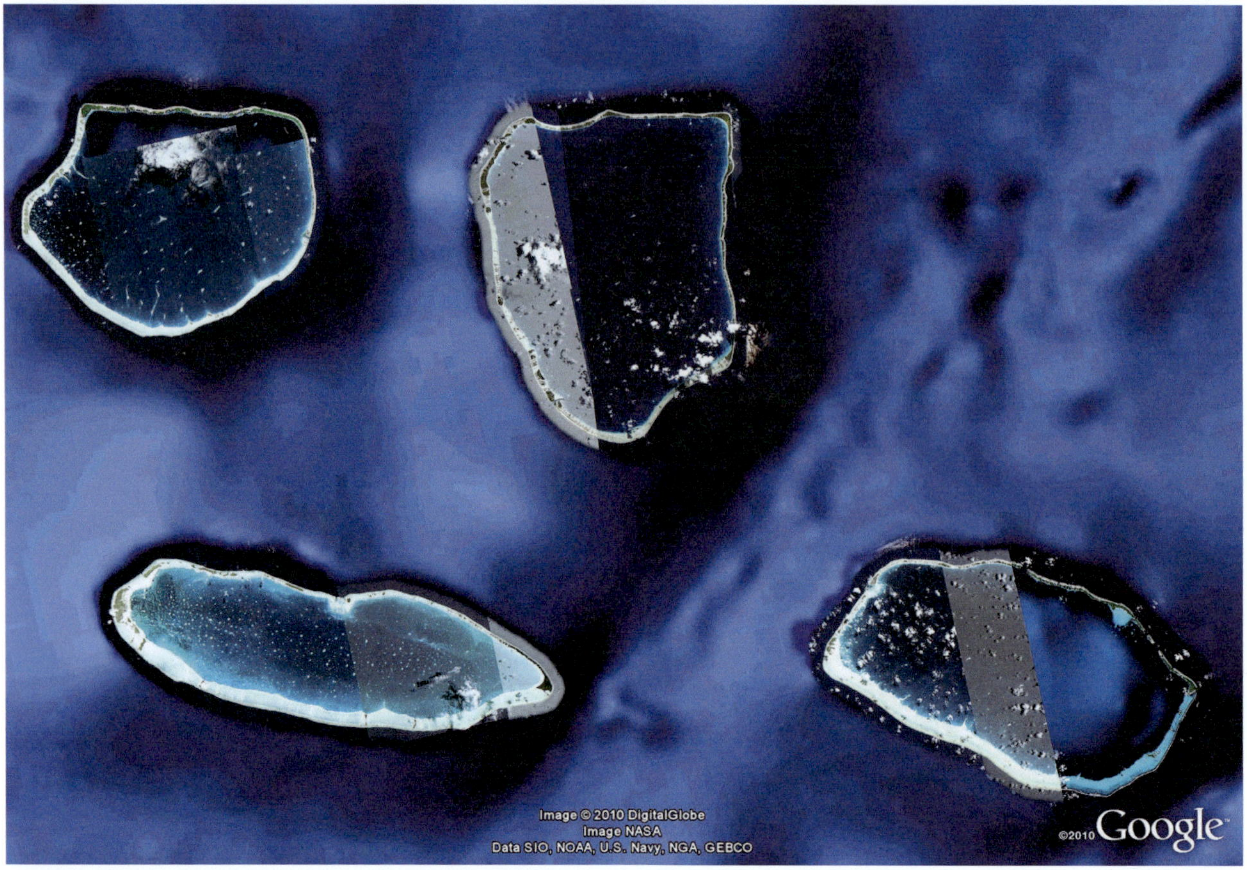
Image © 2010 DigitalGlobe
Image NASA
Data SIO, NOAA, U.S. Navy, NGA, GEBCO
©2010 Google

6.73a

Image © 2010 DigitalGlobe
Data SIO, NOAA, U.S. Navy, NGA, GEBCO
©2010 Google

6.73b

6.4 Other organic hardgrounds

Besides coral reefs, other organisms form "hardgrounds" along coastlines in different (although mostly warm) latitudes, but these structures are too small to be visible on satellite images. These structures are formed by e.g. barnacles, oysters, vermetid gastropods, tubeworms, calcareous algae and other organisms (Baker et al., 2001; Kelletat, 1997, see also Figs. 6.74–6.82). Although tiny and building only thin crusts, it is documented that even along exposed coastlines with strong wave action, organic construction may be dominant over mechanical or bio-erosion/abrasion. Many of these organic hardgrounds occur at specific tidal levels and are valuable past sea-level indicators.

Fig. 6.74 Delicate calcareous algae at a cliff coast of Oregon, USA (Photo credit: D. Kelletat).

◄ Fig. 6.73a,b A group of large atolls (scene is 140 km wide) from French Polynesia in the southern Pacific Ocean, center at about *15° 29′S* and *146° 32′E*. The lower image shows the Mataiva atoll at *14° 53′S* and *148° 40′W*. The atoll has a length of 10 km and a unique internal maze pattern. (Photo credit: ©Google Earth 2010).

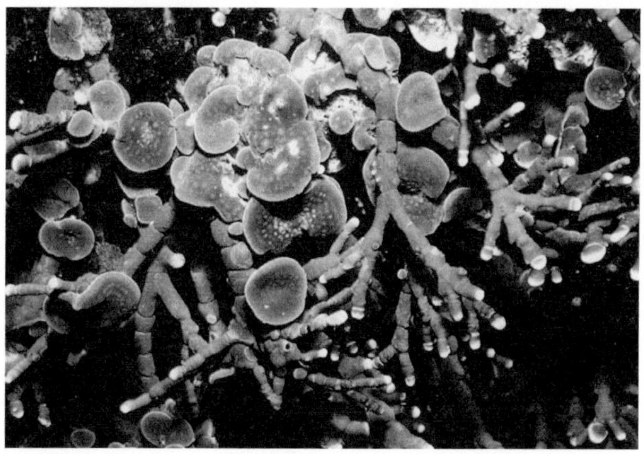

Fig. 6.75 Barnacle rims surround very shallow pools on a rocky platform at Cabo de Trafalgar in southern Atlantic Spain. (Photo credit: D. Kelletat).

Fig.6.76 Banks of rock oysters *(Saccostrea cucullata)* near Quobba, Western Australia, photo taken during low water. (Photo credit: A. Scheffers).

Fig. 6.77 a,b Stromatolites, built by tiny primitive algae, form columnar hardgrounds up to 1 m high in Shark Bay, Western Australia. The species remained almost unchanged since the beginning of organic evolution on Earth. (Photo credit: A. Scheffers).

Fig. 6.78 A broad rim of the vermetid *Dendropoma petraeum* on the west coast of Crete, Greece, developed shortly after the uplift of this part of the island in 365 AD. (Photo credit: D. Kelletat).

Fig. 6.79 Reeflike structures of calcareous algae and vermetids on the west coast of Crete island, Greece, photo taken during high water. Diameter of the forms is up to 10 m. (Photo credit: D. Kelletat).

Fig. 6.80 A micro-atoll 2 m in diameter at low water along the west coast of Crete, Greece. (Photo credit: D. Kelletat).

Fig. 6.81 Rock oyster belt *(Crassostrea amasa)* on a rocky shoreline of Hamilton Island, southern Queensland, Australia. The upper limit of the oysters corresponds to mean high water level. (Photo credit: S. Scheffers).

Fig. 6.82 At many places along limestone coasts bioerosion (the notch) and bioconstruction (the vermetid rim or bench) appear close together, each marking certain environmental conditions (here: duration of seawater exposure). Ibiza Island, Mediterranean Spain. (Photo credit: D. Kelletat).

References

Baker RGV, Haworth RJ, Flood PG (2001) Inter-tidal fixed indicators of former Holocene sea levels in Australia: a summary of sites and a review of methods and models. Quaternary International 83-85:257–273.

Duke NC (2006) Australia´s mangroves. University of Queensland, Brisbane.

Green EP, Short FT (2003) World Atlas of Seagrasses. UNEP-WCMC, Cambridge.

Kelletat D (1987) Das Spektrum biogener Formungsprozesse an den Küsten Neuseelands. Berliner Geographische-Studien 25:421–454.

Kelletat D (1995) Atlas of Coastal Geomorphology and Zonality. Journal of Coastal Research Special Issue No 13.

Kelletat D (1997) Mediterranean coastal biogeomorphology: processes, forms and sea-level indicators. In: Briand F, Maldonado A (ed.) Transformations and evolution of the Mediterranean coastline. CIESM Science Series No 3, Bulletin de l´Institut Océanographique, Monaco Numéro Special 18:209–226.

Short AD, Woodroffe CD (2009) The Coast of Australia. Cambridge University Press, Cambridge.

Spalding MD, Ravilious C, Green EP (2001) World Atlas of Coral Reefs. University of California Press, Berkeley.

Spalding M, Kainuma M, Collins L (2010) World Atlas of Mangroves. Earthscan, GBR.

Steneck RS, Graham MH, Bourque BJ, Corbett D, Erlandson JM, Estes JA, Tegner MJ (2002) Kelp forest ecosystems: biodiversity, stability, resilience and future. Environmental Conservation 29: 436–459.

Veron J (2000) Corals of the world. Australian Institute of Marine Science, Townsville.

Warne K (2011) Let Them Eat Shrimp: The Tragic Disappearance of the Rainforests of the Sea. Island Press, Washington DC.

7

Coasts as Archives of the Past

ABSTRACT What roles have human impacts and natural processes had in shaping the evolution of our world's coastlines during the Holocene? Where, when and how did natural processes such as sea level rise or societies transform the coastal zone? At what scales and rhythms did these changes took place? What can coastal archives tell us about human-environment interactions? Geoarchaeological research attempt to understand the interplay between culture and nature, and more particularly how environments and processes have played a role in Holocene human occupation of the coastal zone. This approach has drawn on the multidisciplinary study of geologic or biologic archives of information, to attempt to differentiate between anthropogenic and natural factors. Other landforms such as uplifted ancient shorelines are evidence for crustal movements in particular in areas of deglaciation and glacio-isostatic uplift. Stepped cliffs, uplifted notches, coastal staircases of ancient coral reefs or fixed biological sea-level indicators allow coastal scientists to reconstruct the history of relative sea-level variations or neotectonics along coastlines.

7.1 Geologic archives in coastal environments

Coasts are very suitable landscapes to analyse environmental changes including tectonics, climatic variations or anthropogenic influences over long time scales as they represent the final sink for sediments near sea level (Anthony, 2009; Baker et al., 2001; Horton et al., 2007; Milne et al., 2009; Pirazzoli 1991, 1996; Sanjaume & Tolgensbakk, 2009; Schwartz, 2005; Brückner & Schellmann, 2003). Excellent sedimentary archives that store these processes – often undisturbed over millennia – in coastal environments are for example old beach deposits, lagoon and marsh sediments, or coastal dunes. Other coastal landforms testify to changes driven by tectonic forces within the Earth. Uplifted coastal cliffs or coral reef terraces high above modern sea level are expressions of tectonic uplift whereas drowned coastal features or ancient archaeological remains may document subsidence of the coastal region (when sea level changes are excluded as a cause factor). The best circumstances for the preservation of past coastlines often can be found in areas of uplift, where forms and deposits are moved out of the influence of wave attack.

Steps in a coastal environment (either sedimentary as in Fig. 7.1 from the staircase of coastal terraces along Metaponto Bay in southern Italy, or as rocky features like the uplifted inactive cliffs of Peru and Chile, Figs. 7.2 and 7.3) allow to have a look into ancient surf belts or even shallow sea floors. In studying coastal terraces built by coral reefs as on the Huon Peninsula of Papua-New Guinea, Haiti (Figs. 7.4 and 7.5) or Cuba (Fig. 7.6), coastal scientists can reconstruct past climatic and oceanographic conditions via the identifica-

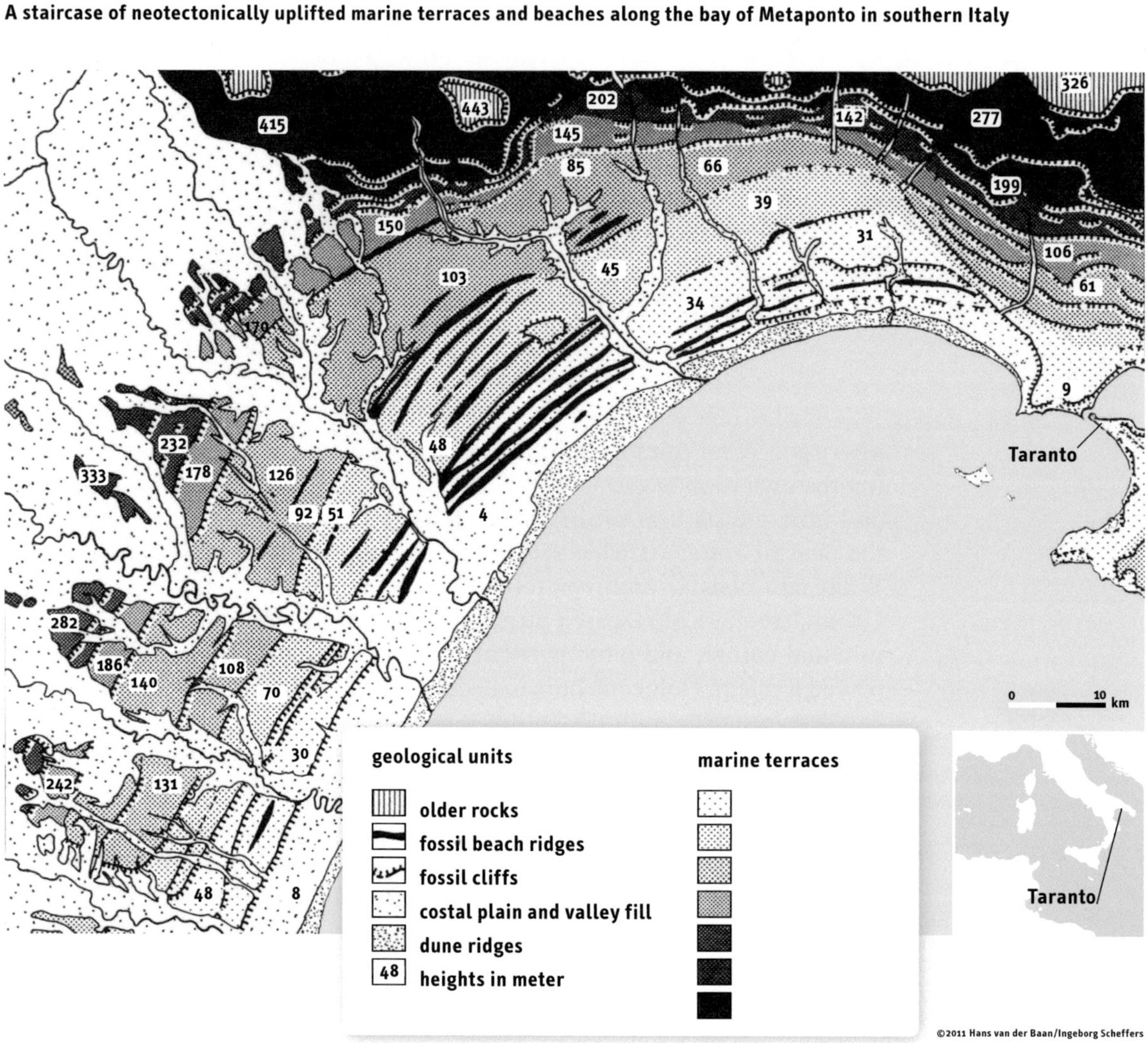

Fig. 7.1 A staircase of tectonically uplifted marine terraces and beaches along the bay of Metaponto in southern Italy. (Photo credit: H. Brückner).

tion of coral communities or apply geochemical methods to extract information about sea surface temperatures or water quality from the carbonate skeletons of fossil corals. In ancient sedimentary coastal environments, old shorelines as beaches or beach ridges may be preserved and uplifted out of the reach of wave attack as is visible in well developed examples along the Peruvian and Chilean coastlines (Figs. 7.7 and 7.8). Sequence of beach ridges in glacio-isostatic uplifted arctic areas exist inland for many kilometres and can be followed (as in Scandinavia or northern Canada) up to 300 m above present sea level (Fig. 7.9). In other areas of strong uplift, notches or former rocky shore platforms may be preserved in higher altitudes, and sometimes remnants of littoral and marine sediments, including mollusc or marine microfossils suitable for isotopic dating are found which enables geochronologist to establish a good understanding of when these changes took place in the past (see also Chapter 1).

For almost 100 years earth scientists have studied a wide diversity of marine organisms that live in the foreshore zone as indicators of past sea levels including certain foraminifera com-

►Fig. 7.2 Inactive cliffs due to tectonic uplift along the coastline of Peru at *15° 05′S* and *75° 23′W*, width of the image of about 16 km. (Photo credit: ©Google Earth 2010).

►Fig. 7.3 At least three steps 10 km inland resulting from tectonic uplift can be seen along this section of the Chilean coastline at *30° 51′S* and *71° 39′W*. (Photo credit: ©Google Earth 2010).

Image © 2010 GeoEye
Data SIO, NOAA, U.S. Navy, NGA, GEBCO
Image © 2010 TerraMetrics
©2010 Google

7.2

Image © 2010 TerraMetrics
Data SIO, NOAA, U.S. Navy, NGA, GEBCO
Image © 2010 DigitalGlobe
©2010 Google

7.3

Image © 2010 GeoEye
Image © 2010 TerraMetrics
Data SIO, NOAA, U.S. Navy, NGA, GEBCO
Image © 2010 DigitalGlobe
©2010 Google

7.4

Image © 2010 DigitalGlobe
© 2010 Cnes/Spot Image
Data SIO, NOAA, U.S. Navy, NGA, GEBCO
Image © 2010 GeoEye
Google

7.5

Fig. 7.6 Along the south coast of Cuba up to 13 elevated coral reef terraces document ongoing uplift processes. A 15 km wide section at about *19° 51′N* and *77° 31′W*. (Photo credit: ©Google Earth 2010).

munities in coastal sediments, corals, borings of marine bivalves or sea urchins exposed over low water or oysters, barnacles or tubeworms out of the reach of mean sea level and the surf and spray zone (Fig. 7.10 and 7.11). Fossil examples of these marine organisms can reveal past changes of relative sea level that – with the help of isotope geochronology such as radiocarbon or U-series dating – can be reconstructed with a precision of centimetres.

These ever changing isostatic, eustatic or tectonic forces can not only change the location of the sea relative to the land but also make finding palaeo-shorelines with coastal archaeological sites a tricky business as we see in the next section.

◀ Fig. 7.4 A staircase of coral reef terraces from the Pleistocene at the north coast of Haiti at *19° 50′N* and *73° 16′W*. Width of scene is about 21 km. (Photo credit: ©Google Earth 2010).

◀ Fig. 7.5 Detail from Fig. 7.4. (Photo credit: ©Google Earth 2010).

Image © 2010 Ge

Fig. 7.7 Delicate forms of beach ridges up to about 100 m asl and 13 km inland identify a rapid rise by tectonics at the Chilean coast at *27° 15′S* and *70° 54′W*. (Photo credit: ©Google Earth 2010).

Image © 2010 GeoEye
Data SIO, NOAA, U.S. Navy, NGA, GEBCO
©2010 Google

7.8

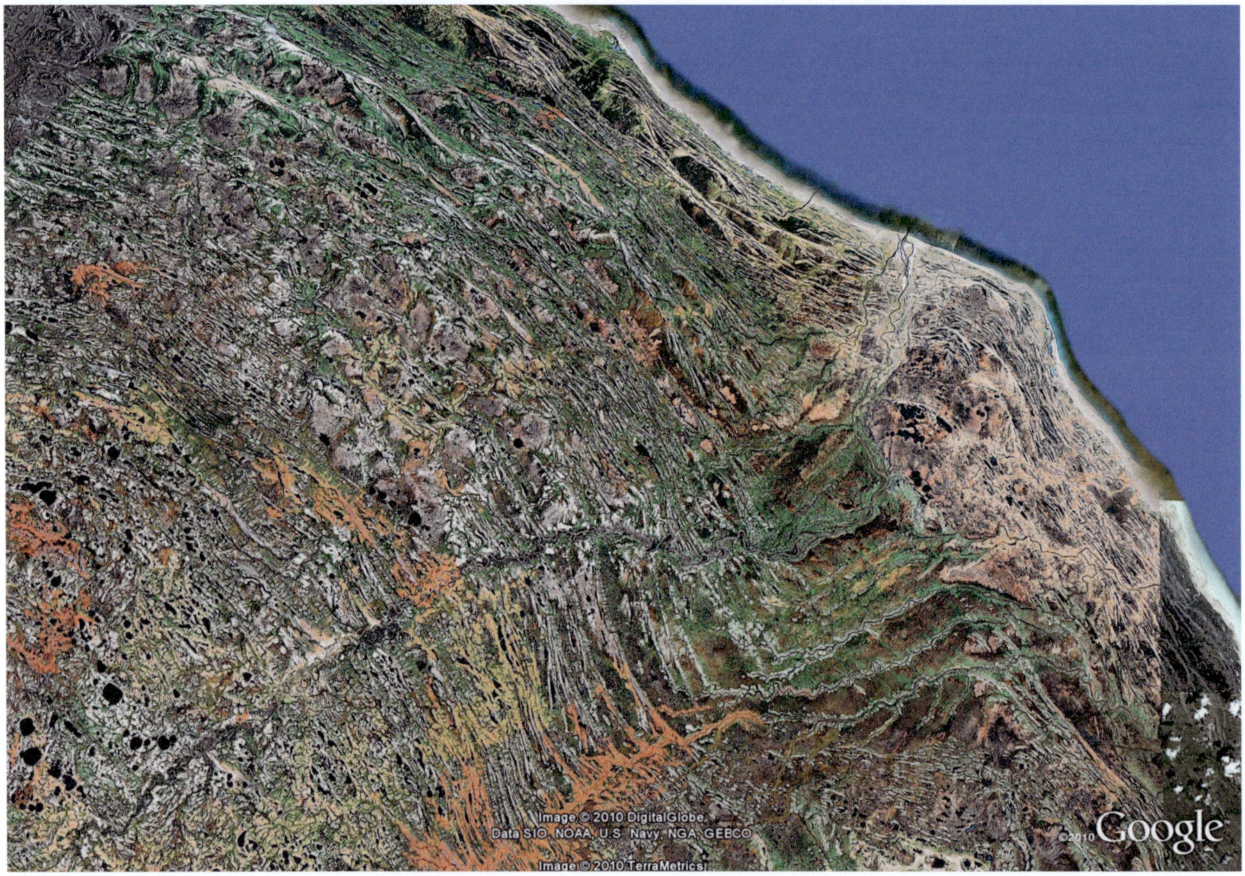
Image © 2010 DigitalGlobe
Data SIO, NOAA, U.S. Navy, NGA, GEBCO
Image © 2010 TerraMetrics
©2010 Google

7.9

Fig. 7.10 Benches of rock oysters of the species Saccostrea cucullata in southern Thailand are excellent "fixed biological sea-level indicators", here for the MHW-level at the upper limits of growth. (Photo credit: D. Kelletat).

Fig. 7.11 a Coral microatolls (here: Porites lutea) document exact mean lower sea levels for the identification of modern and past tidal and sea levels. (Photo credit: D. Kelletat).

◄ Fig. 7.8 Another example of parallel beach ridges elevated by tectonism and 13 km inland along the Peruvian coast at *15° 37′S* and *74° 35′W.* (Photo credit: ©Google Earth 2010).

◄ Fig. 7.9 Numerous (>100) beach ridges elevated by glacio-isostatic rebound of the lithosphere after deglaciation along the shoreline of Hudson Bay, Canada, at *56° 06′N* and *88° 37′W,* reaching at least 65 km inland. (Photo credit: ©Google Earth 2010).

Fig. 7.11b Coral microatolls (here: Porites lutea) document exact mean lower sea levels for the identification of modern and past tidal and sea levels. (Photo credit: D. Kelletat).

7.2 Coastal Geoarchaeology

Humans have inhabited coastal regions and accessed their rich overlapping maritime, littoral and inland resources for hundreds of thousands of years. But we have seen that throughout Earth's history the boundary between land and sea was ever-changing and thus, sites of ancient coastal societies may be well inland, well submerged under the sea or still coastal. For example, prehistoric coastal settlements of the Younger Palaeolithic or Mesolithic period are mostly located offshore and covered today by up to 120 m of water as the rising postglacial sea level flooded the fringes of the continents and shelf regions. Thus, many prehistoric sites of these ancient cultural epochs have been lost and subject of underwater archaeology. It was only at the end of the Holocene transgression, around 6.000 years ago, that our ancestors started to settle along "modern" coastlines as we see them today.

While archaeologists most often use the results of other scientists to help interpret their sites, occasionally archaeological evidence is use to help interpret the geologic and environmental history of a landscape (Johnson, 1995). Most of these studies are related to the understanding of past sea levels as people do not live underwater and thus sites found beneath sea level are documenting a rising sea level or subsidence of the land (Figs. 7.12–7.16). On the other hand, sites with coastal or maritime infrastructure located well away from the shore indicating lower sea level or coastal uplift. In trying to understand how ancient societies lived and adapted to a specific coastal zone in the past and responded to short- and long-term environmental changes, coastal geoarchaeologists apply a multidisciplinary approach which not only includes archaeology but also geology, geomorphology, geography, history or even marine biology.

One great example of how archaeologistst and geoscientists work together to reconstruct environmental changes and human response is the ancient Greek city of Miletus on the western coast

Fig. 7.12 Shell middens are anthropogenic deposits which contain numerous amounts of shells and can be found from subarctic to tropical latitudes in coastal areas and along rivers or lakes where freshwater molluscs have been collected as a diet by people. The Neolithic settlement of Skara Brae on the Orkney Islands north of Scotland, is one of the oldest settlement preserved in Europe. The associated shell middens document that the village was situated along the shoreline more than 5.000 years ago – in a position similar to the modern coast. The settlement was hidden under a coastal dune for several thousand years and only discovered again after a storm in the 19th century. (Photo credit: D. Kelletat).

of Anatolia in Turkey (Fig. 7.19). The history of the landscape was reconstructed through the geological and geoarchaeological evaluation of the alluvial plain and delta-archives. More than one hundred sediment cores were taken by means of a percussion-coring device, followed by sedimentological, petrological and palaeoecological examination in the laboratory. The earliest available archaeological evidence documents that the islands on which Miletus was originally placed were inhabited by a Neolithic population around 3500–3000 BCE. In the early and middle Bronze age the settlement came under the influence of Greek Minoans from Crete and the site was renamed Miletus after a place in Crete. During that time Miletus was considered the greatest and wealthiest of Greek cities. The city possessed a harbour at the southern entry of a large bay, but during the last six millennia, the former marine embayment of the

Fig. 7.13 The submerged cart tracks between the mainland of southern Greece and Elaphonisos Island are documents that the modern water way between the mainland and the island did not exists in ancient times. The tracks belong to the remnants of a Helladic settlement which flourished 3.500 years ago. Whether the tracks are drowned because of sea level rise or tectonic subsidence is still an open question. (Photo credit: A. Scheffers).

Fig. 7.14 A partly drowned Minoan grave at Malia at the north coast of Crete Island, Greece, documents a relative subsidence or sea level rise during the last 3.500 years of at least 2.5 m at this place. (Photo credit: A. Scheffers).

Fig. 7.15 Roman ruins submerged in the water on the north coast of Crete. (Photo credit: D. Kelletat).

Latmian Gulf has been silted up by the progradation of the Maeander (Büyük Menderes) Delta. Today, the ruins of Miletus are an inland location as long-term human impact together with an ecologically unstable natural environment in the Mediterranean has led to strong erosion in the hinterland and the resulting delta progradation and gradual infill of the embayment. The historic delta growth (of around 30 km) of the Maeander River is one of the most spectacular cases of delta progradation in the Mediterranean region. The westward shift in the shoreline has been documented in the ancient literature (e.g. Herodotus, Strabo, Pausanias), by archaeological evidence from the former seaport cities Miletus, Priene, Myous and Herakleia, and by palaeogeographical studies (Brückner 2003; Müllenhoff 2005).

Due to the progressive postglacial sea level rise, lower parts of the Late Neolithic settlement were flooded and people had to move to higher grounds. It is concluded that relative sea level was highest during Early and Middle Bronze Age (3.000–2.000 BC) when the transgression created an archipelago-like coastal landscape. This is confirmed by the Holocene sea level curve for the whole of the lower Büyük Menderes Delta and floodplain since it also peaks around 2.500 BC (Müllenhoff 2005). Archaeological remains indicate that at 1.700 BC it became possible to resettle the area. This implies that in the meantime rela-

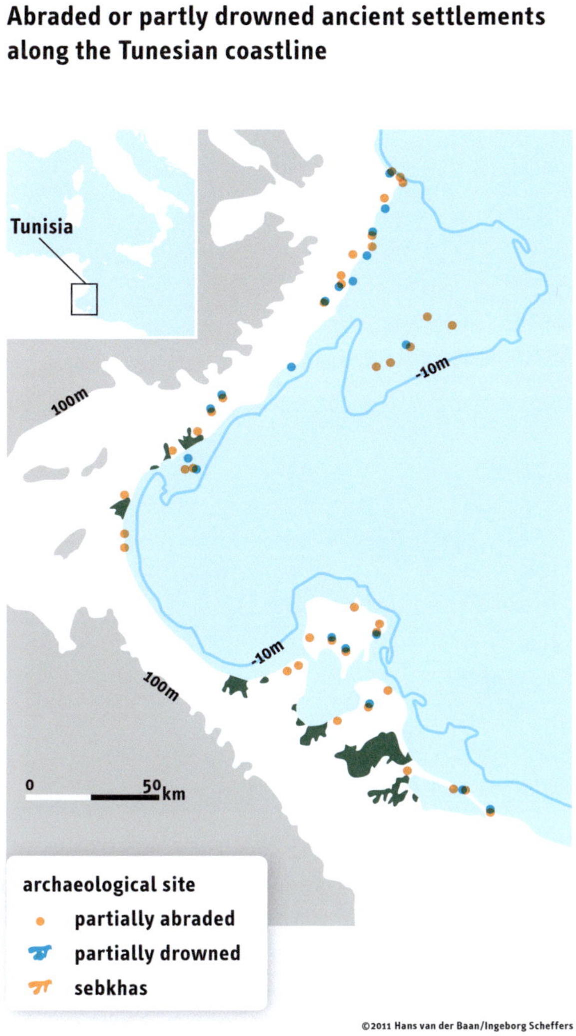

Fig. 7.16 Eroded or partly drowned ancient settlements can be found along a part of the Tunisian coastline (acc. to Paskoff & Oueslati, 1991).

Fig. 7.17 The fundaments of a Viking boat shed (9th–10th century AD) on the Shetland Islands of northernmost Scotland indicates that the position of the shoreline has not changed during the last 1.000 years. (Photo credit: D. Kelletat).

Fig. 7.18 The ruins of Antrims Castle near Giant's Causeway in Northern Ireland which was built in Medieval times on a steep promontory. (Photo credit: D. Kelletat).

Büyük Menderes Delta Progradation, since 1500 BC

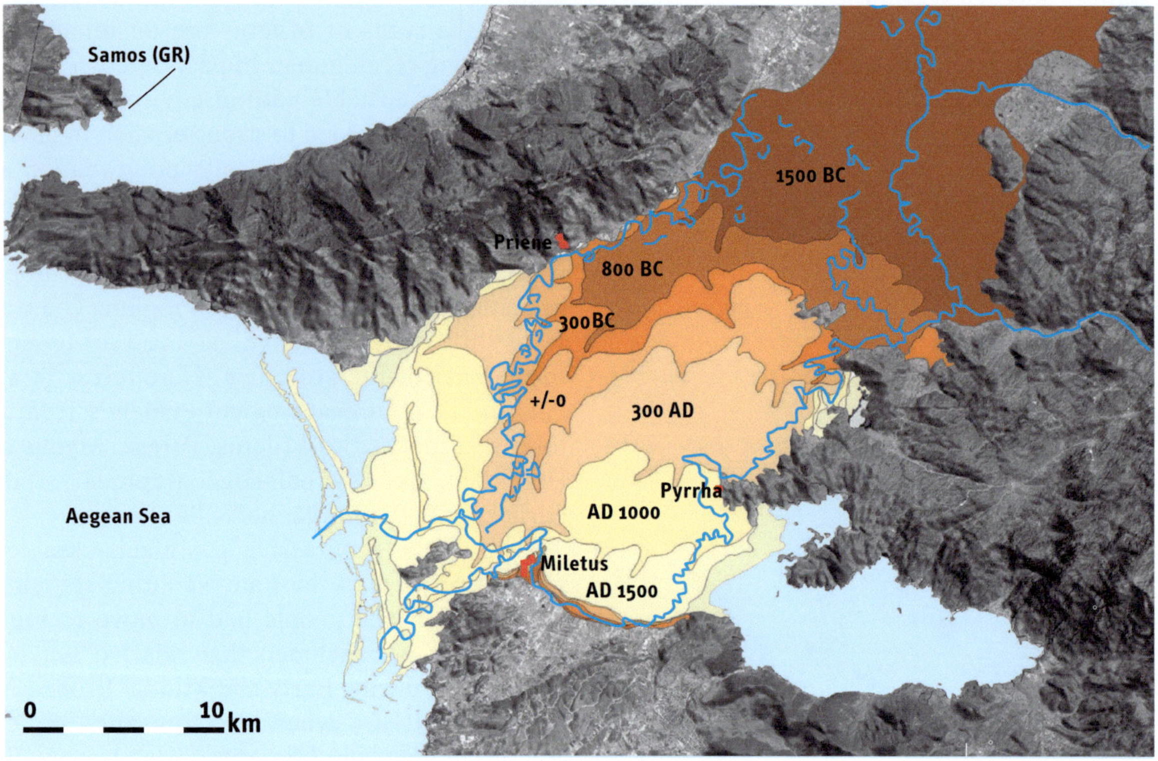

Fig. 7.19 The palaeo-geographic evolution of the Maeander delta shows an aspect about 3.500 years ago, when the modern delta did not exist and Miletus was situated on a rocky island and had several harbours. (Image credit: Brückner et al., 2005).

tive sea level had fallen which is also shown by regressive and littoral sediments encountered in the cores. Due to this regression, denudation processes and coastal dynamics, the archipelago subsequently turned to the Milesian peninsula during the 2nd millennium BC. In Roman Imperial times, the peninsula became landlocked by the prograding Maeander Delta, because sedimentation rates were especially high triggered by intensive land use, clearing of forests, and livestock farming.

References

Anthony EJ, (2009) Shore processes and their palaeoenvironmental applications. Marine Geology 4:264–288, 319–324.

Baker RGV, Haworth RJ, Flood PG (2001) Inter-tidal fixed indicators of former Holocene sea levels in Australia: a summary of sites and a review of methods and models. Quaternary International 83–85:257–273.

Brückner H, Schellmann G (2003) Late Pleistocene and Holocene shorelines of Andréeland (Spitsbergen, Svalbard) – Geomorphological evidences and palaeo-oceanographic significance. Journal of Coastal Research Coast Res 19 (4):971–982.

Brückner H (2003) Delta evolution and culture – aspects of geoarchaeological research in Miletos and Priene. In: Wagner GA, Pernicka E & Uerpmann HP (eds.): Troia and the Troad. Scientific approaches. – pp. 121–144. Springer Series: Natural Science in Archaeology. Berlin, Heidelberg, New York.

Brückner H, Muellenhoff M, Gehrels R, Herda A, Knipping M, Voett A (2006) From archipelago to floodplain – geographical and ecological changes in Miletus and its environs during the past six millennia (Western Anatolia, Turkey). Annals of Geomorphology, N. F., Suppl. – Vol. 142: 63–83.

Horton BP, Zong Y, Hillier C, Engelhart S (2007) Diatoms from Indonesian mangroves and their suitability as sea-level indicators for tropical environments. Marine Micropaleontology 63:155–168.

Johnson, L (2005) Archaeology. In: Schwartz ML (Ed., 2005): Encyclopedia of Coastal Science. Springer, Dordrecht.

Milne GA, Gehrels WR, Hughes CW, Tamisiea ME (2009) Identifying the causes of sea-level change. Nature Geoscience Vol 2:471– 478.

Pirazzoli PA (1991) World Atlas of Holocene Sea-Level Changes. The Netherlands: Elsevier Oceanography Series 58, Amsterdam.

Pirazzoli PA (1996) Sea-level changes: the last 20 000 years. Wiley, Chichester.

Sanjaume E, Tolgensbakk J (2009) Beach ridges from the Varanger Peninsula (Arctic Norwegian coast): Characteristics and significance. Geomorphology 104 (1–2):82–92.

Schwartz ML (Ed., 2005): Encyclopedia of Coastal Science. Springer, Dordrecht.

8

Coasts at Risk

ABSTRACT Among all observed natural hazards, water-related disasters are undoubtedly the most recurrent and pose major threats to human security and sustainable socio-economic development, as recently witnessed with the disasters caused by the Japan tsunami in 2011, the Indian Ocean tsunami in 2004, Hurricane Katrina in 2005, Cyclone Nargis in 2008, and many others. Ever since first settling along the coast, human societies had to adapt to the constantly changing conditions and to the risk of storms and floods. Today, at a rough estimate, more than 200 million people worldwide live in coastal areas less than 5 meters above sea level. By the end of the 21st century the United Nations University Institute for Environment and Human Security (UNU-EHS) estimates this figure to increase to 400–500 million as coastal megacities grow rapidly in most countries. Already during the period 2000 to 2006, a total of 2.163 hydrological disasters were reported globally in the Emergency Disasters Database (EM-DAT), killing more than 290.000 people, affecting more than 1.5 billion, and inflicting more than US$422 billion of damage. Coastal areas are among the most densely populated regions of the world and therefore particular vulnerable to the impacts of storms and tsunamis and the impacts of climate change such as sea level rise and associated coastal erosion. Researchers all around the world seeking answers to the hotly debated million dollar question of how rapidly and to what extent sea level will rise as a consequence of climate change. On old saying from medieval times simply states "Build a dyke or move away", today policymakers and governments are facing challenging discussions of the future of our coasts – defence or orderly retreat.

8.1 Coastal Natural Hazards – Storms and Tsunamis

Coastlines are shaped by natural forces but few things in nature can compare with the destructive force of a cyclone or a tsunami. In recent years the tsunamis in the Indian Ocean (2004), Samoa (2009), Chile (2010), Japan (2011) and extreme hurricanes as Katrina (2005), Ike (2008) or Mitch (1998) not only brought total devastation, but they also brought the way of living too close to the shoreline – something that is very much a reality for coast-dwellers in developed and less-developed nations alike – to a sudden and traumatizing end. At a rough estimate more than 200 million people worldwide live along coastlines less than 5 metres above sea level and thus are vulnerable to marine hazards. By the end of the 21st century this figure is estimated to increase to 400 to 500 million. The 2005 Atlantic hurricane season that included was the most catastrophic ever recorded, and it changed the way the insurance industry conducts business. The first quarter of 2011 already is the most loss-afflicted in reinsurance history in terms of natural catastrophes: Cyclone Yasi causing widespread flooding in eastern Australia, a second strong earthquake in New Zealand after the

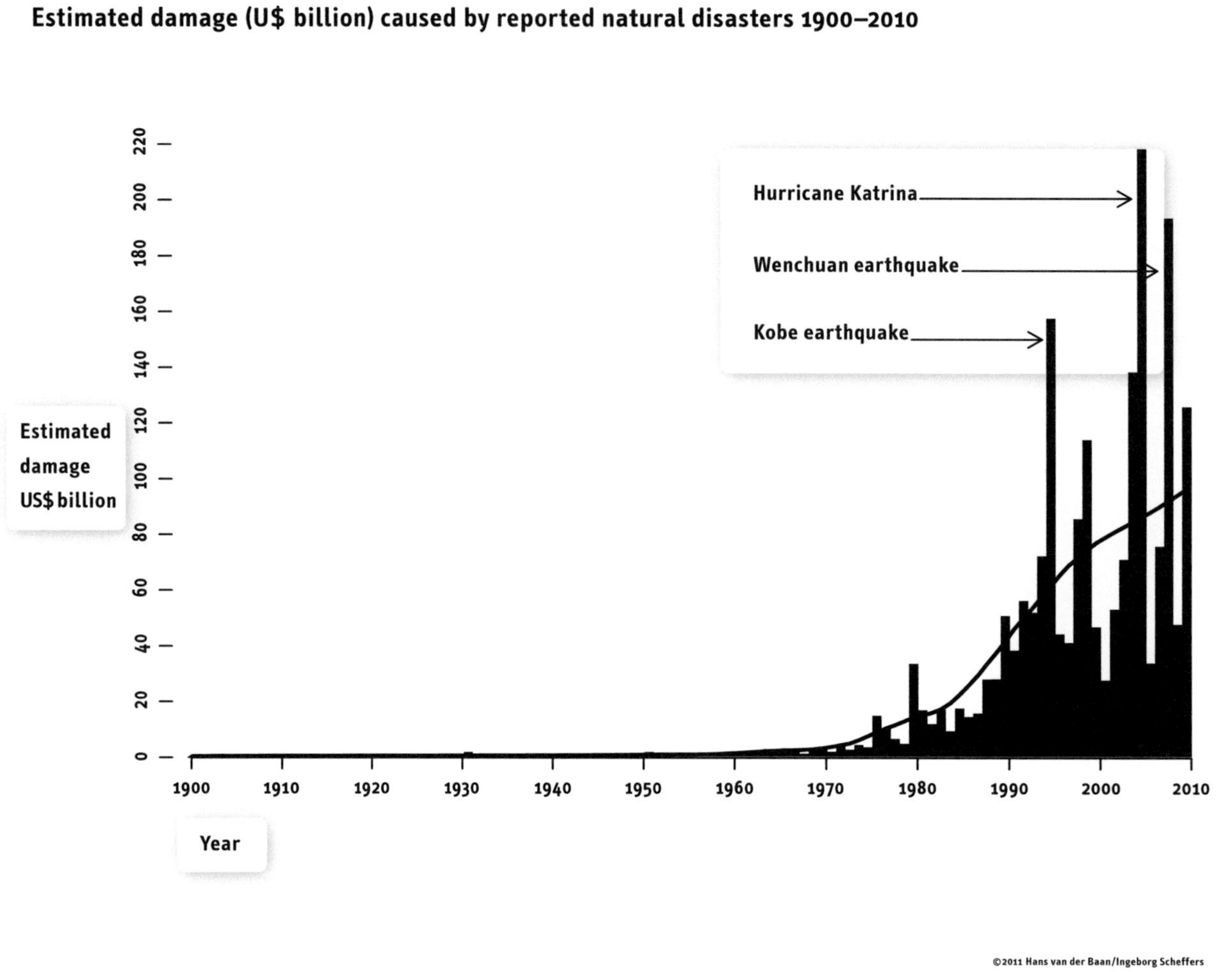

Fig. 8.1a Since 1988 the WHO Collaborating Centre for Research on the Epidemiology of Disasters (CRED) has been maintaining an Emergency Events Database EM-DAT. The main objective of the database is to serve the purposes of humanitarian action at national and international levels, but is a rich source for the interested public as well. Estimated damage (US$ billion) caused by reported natural disasters 1900–2010.

disastrous Christchurch earthquake in 2010, and then the devastating earthquake in Japan. With an intensity of 9.0, it was the strongest quake ever recorded in Japan and the fourth-severest ever measured anywhere in the world. The world's greatest insurance company for natural disaster, the Munich Reinsurance Company, expect this event alone to give rise to a claims burden of $210 billion US dollar. Faced with potential staggering costs and the prospect of continuing global climate changes, insurers began abandoning coastal markets (Fig. 8.1, 8.2). Yet coastal population density continues to rise! In the next short sections we briefly discuss tsunamis and storm surges, both very complex events that consist of many different components that influence their effect on the coastal zone.

Storms and storm surges

Regarded as the greatest storms on Earth, cyclones (synonym to hurricanes) can cause havoc or in some cases completely annihilate shorelines and coastal regions with sustained winds, lasting many hours, of 260 km/hour or higher combined with intense rainfall and a storm *surge*. Earth scientists have estimated that during its life cycle, a hurricane can expend as much energy as 10.000 nuclear bombs! The term *hurricane* originates from Huracan, a god of evil to the Indigenous people of Central America. In the western Pacific and China Sea, hurricanes are known as *typhoons*, from the Cantonese word tai-fung, meaning "great wind." In the north-western Pacific, Bangladesh, Pakistan, and India, they are known as *cyclones*; in the Philippines, they are called *baguios*.

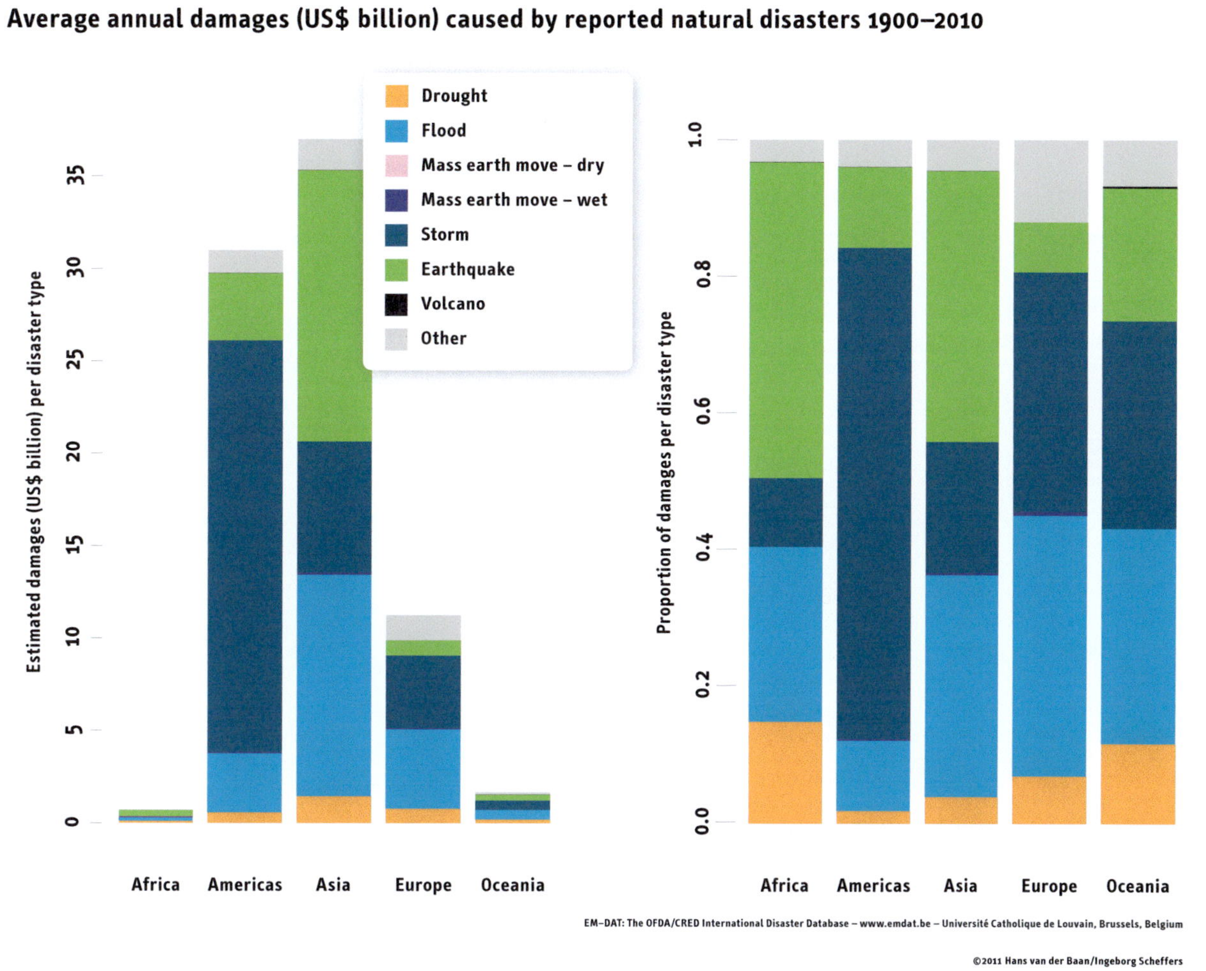

Fig. 8.1b Since 1988 the WHO Collaborating Centre for Research on the Epidemiology of Disasters (CRED) has been maintaining an Emergency Events Database EM-DAT. The main objective of the database is to serve the purposes of humanitarian action at national and international levels, but is a rich source for the interested public as well. b) Average annual damages caused by reported natural disasters 1990–2010. (Source: EM-DAT, OFDA/CRED International Disaster Database – www.emdat.be).

Hurricanes form over tropical parts of Earth's oceans, between 8° and 20° latitude, in areas of high humidity, light winds, and warm sea-surface temperatures (typically 26°C or greater). These conditions usually occur in the summer and early fall months of the tropical North Atlantic and North Pacific oceans, with hurricane events sharply peaking from late August through September. In the southern hemisphere, the cyclone season extends in the summer month from 1 November through to 30 April. The first sign of hurricane development is the appearance of a cluster of thunderstorms over the tropical oceans in a region where tropical winds converge on one another. Occasionally, a cluster of thunderstorms breaks out from this convergence zone and becomes better organized. As the storm develops, water vapour condenses to form rain, which releases heat energy. In response to atmospheric heating, the surrounding air becomes less dense and begins to rise. The atmospheric pressure at sea level drops in the region of heating. As the warm air rises, it triggers more condensation and rainfall, which in turn releases more heat. This causes more air to rise, which causes the surface pressure at sea level to fall even further. At this point a chain reaction occurs, as the rising temperatures in the centre of the storm cause surface temperatures to fall to progressively lower levels. In the Northern Hemisphere, because of the Coriolis effect, the increasing winds will begin to circulate in a counter-clockwise pattern around the storm's area of lowest pressure, which becomes the "eye" of the hurricane, on

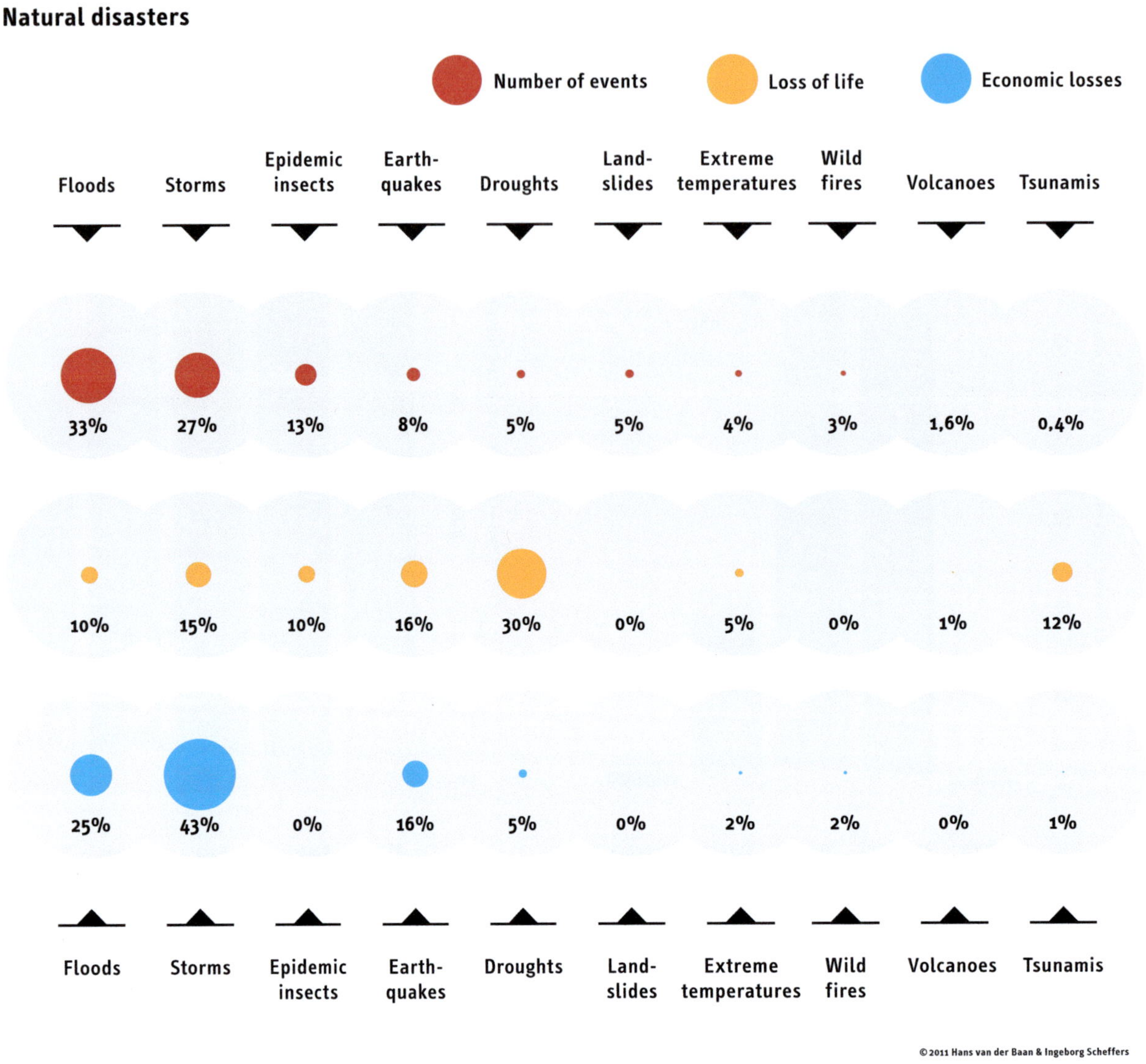

Fig. 8.1c Reported disaster types, economic damages and loss of life from natural disasters in 1960- 2000. (Source: EM-DAT, International Disaster Database).

the southern hemisphere the cyclones circulate clockwise. Hurricanes or tropical cyclones are classified into three main groups, based on intensity: Once sustained wind speeds reach 37 km/hour, the storm system is called a tropical depression. As winds increase to 63 km/hour, the system is called a tropical storm and receives a name. This naming tradition started with the use of World War II code names, such as Andrew, Bonnie, Charlie, and so forth. Finally, when wind speeds reach 119 km/hour, the storm is classified as a hurricane. Once it becomes a hurricane, the storm is assigned a 1–5 rating based on its current intensity. This hurricane *intensity scale* (Saffir-Simpson scale) is used to estimate the potential property damage and flooding expected along the coast from hurricane landfall.

On August 29, 2005, Hurricane Katrina made landfall in Louisiana and struck New Orleans as a nearly category 4 storm, with sustained winds of 204 km/hour (127 miles/hour). Later that day, the rise in sea level associated with the hurricane storm *surge* caused several sections of the levee system in New Orleans to collapse. Subsequent flooding of parts of the city to depths of up to 7 or 8 m claimed hundreds of lives and left the city submerged and abandoned for almost a month. The reverberations are still felt in the lives of many people coming back to revive New Orleans. The disaster in New Orleans was not so much the

Fig. 8.2 Miami Beach in east Florida with a skyscraper skyline very close to a narrow beach with frequent strong hurricane impacts. (Photo credit: ©Google Earth 2010).

Fig. 8.3 At times of especially high water levels (during spring tides or storm surges) in the Lagoon of Venice some areas of the city, like the Piazza San Marco, are repeatedly flooded. Italians call the high water "acqua alta". (Photo credit: Istockphoto LP).

◀ Fig. 8.4a The series of photographs before and after show how these extreme storm events may change landscapes within a decade. The top image was taken in July 2001, before Hurricane Lili (2002). The middle photograph was taken on September 17, 2004, immediately after the passage of Hurricane Ivan. The bottom image was acquired on August 31, 2005, two days after Hurricane Katrina. Note the road and two parallel canals in the first photograph. The post-Ivan photo shows overwash deposits covering the road and migrating towards the first canal. The post-Katrina photo shows that the overwash deposit has not only covered the road but also filled the first canal and is approaching the second. The beach appears brown in the bottom photograph because of a "deposit" of plant debris. (Photo credit: NASA).

Fig. 8.4b Dramatic shoreline erosion and large overwash deposits along Dauphin Island during Hurricane Katrina (2005) demonstrate classic barrier island rollover. (Photo credit: USGS)

direct impact of the hurricane itself as of the storm *surge* that ultimately caused the collapse of part of the city's levee system.

The damaging effects of extremely high, sustained winds and torrential rains are intuitively easy to understand and may make up the majority of headlines as a hurricane bowls its way across the ocean towards cities, but often the associated storm *surge*, in which major regions of the coastline become flooded, is potentially the most destructive influence of these epic storms.

As a hurricane moves across the ocean, the high winds and low pressure created by the storm act like a straw and trigger a hump of water at the centre of the storm – the storm surge. Many complicating factors can influence a storm surge such as tides, the water depth off the coastline or the bathymetry (or topographical features) of the ocean bottom. The strong winds inside the hurricane act like a plow, causing water to "pile-up" along the front of the storm, particularly along the right-front quadrant where also the highest wind velocities are recorded. These two effects cause a large bulge of water to develop. Over deep water, far from land, this water bulge is allowed to flow away, keeping the rise in sea level small. As the storm moves closer to the coastline into shallow waters, the water has "nowhere to go" and the mound of water grows. When the hurricane moves onshore, and particularly if landfall coincides with high tide, vast quantities of water are amassed along the coastline and flood large areas of land. High waves may further intensify this situation.

Where low-lying coastlines are protected by dikes, seawater cannot flow back into the sea after flooding has occurred. Furthermore, a storm surge will push up a river estuary and can cause damage over great distances inland. Storm surge moves with the forward speed of the hurricane typically 15–25 km/h. Imaging the destructive power of storm surge a one cubic meter of sea water weighs on average 1.027 kg – almost a ton. Compounding to the rushing water is the large amount of floating debris that typically accompanies the surge. Trees, pieces of buildings and other debris float on top of the storm surge and act as projectiles that can impact on buildings or other structures. The storm surge can begin to rise a day before the storm hits and affect areas which are not in the direct path of the hurricane.

Recent hurricane impacts like Katrina (2005), Hurricane Ike in 2008, the third-costliest hurricane ever to make landfall in the United States and cyclone events in other ocean basins, together with growing populations have made clear the need for a more accurate and extensive record of storm activity. But in most countries the observational records are not long enough and the paucity of historical records of cyclone events limits our understanding of the temporal and spatial

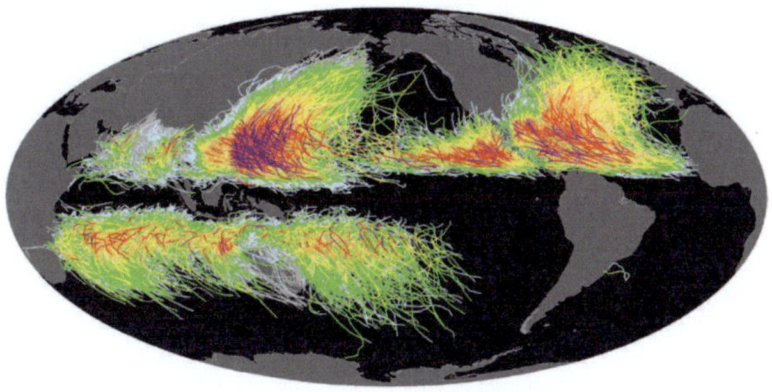

Fig. 8.5 Tracks of all tropical cyclones in the world's ocean basins. The colours represent the hurricane categories: from green (cat 1 = about 125 km/h sustained winds) to dark red (cat. 5 = more than 250 km/h sustained winds). Given the significant impacts of land-falling cyclones, much research has gone into understanding these tremendous forces of nature. IBTrACS stores information for all known tropical cyclones recorded since the 1850s. In the 19th century, mariners who encountered tropical cyclones recorded the storms' locations in ships' logbooks. The only storms recorded during that century were those that struck inhabited land or were noted in surviving logbooks. Today, virtually no storm goes undetected owing to the nearly continuous satellite coverage of our entire planet. (Image credit: NOAA; Access under: http://www.climatewatch.noaa.gov/2010/images/tropical-cyclone-tracks-throughout-history/http://www.ncdc.noaa.gov/oa/ibtracs/index.php).

Fig. 8.6 The Rosenberg Library's presentation of the Galveston Storm of 1900 in Google Earth visualises the impact of the deadliest storm in the US history. The category 4 hurricane made landfall on the city of Galveston in the U.S. state of Texas, on September 8, 1900. The Rosenberg Library's Google Earth project lends geographic organization to the large collection of photographs that chronicle how a 15.7 ft high storm surge devastated an island no more than 8.7 ft above sea level at its highest point, destroying 3,600 buildings and killing 6,000 to 8,000 people in the process. More importantly, this project chronicles the incredible will of Galvestonians to protect the island from future storms by building a 17 ft tall seawall, jacking up 2,000 buildings on stilts, and raising the ground level of 500 city blocks. Read more about it and download the Google Earth application at http://sites.google.com/site/galvestonstorm/ (Photo credit: ©Google Earth 2010).

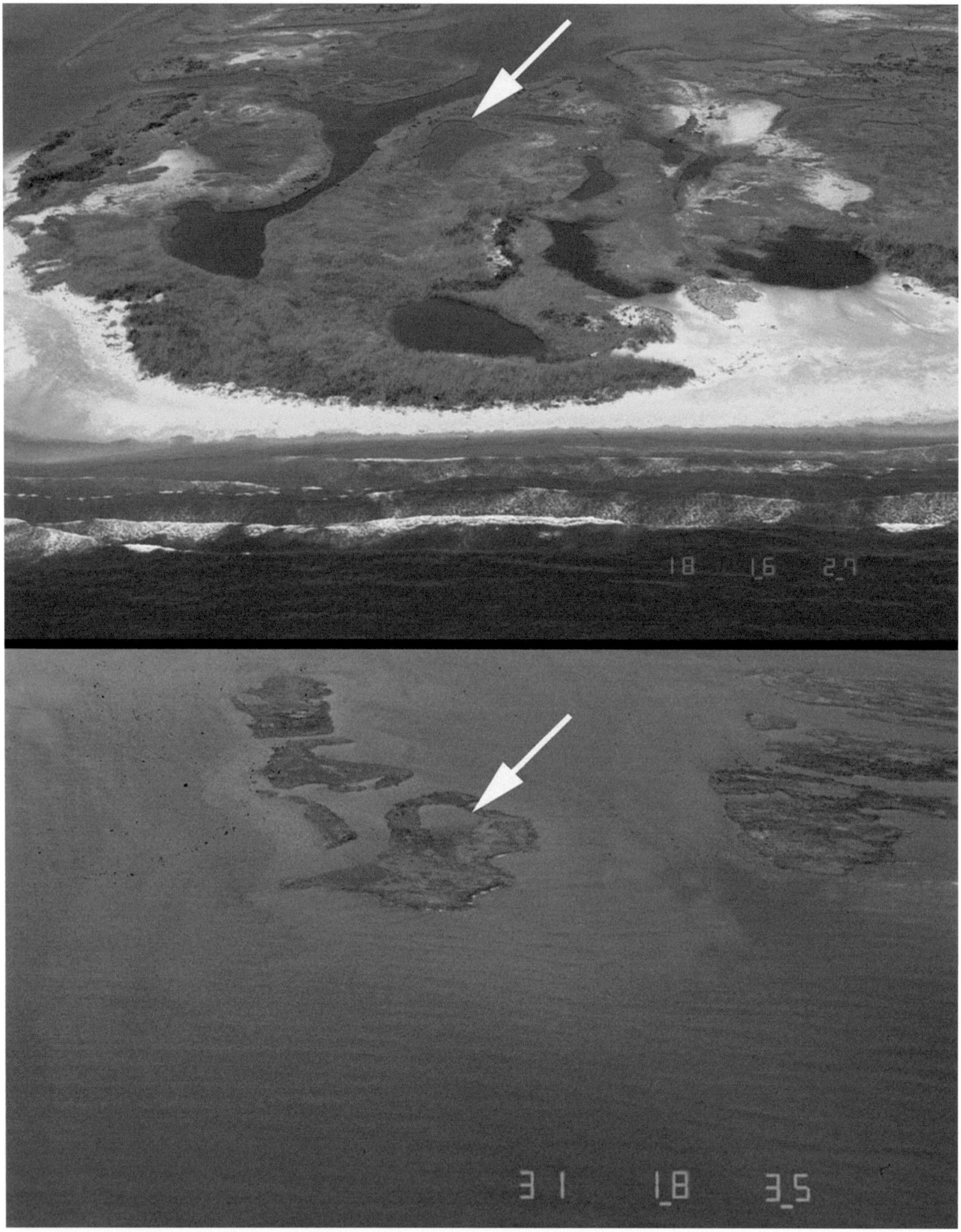

Fig. 8.7 Northern Chandeleur Islands. The first image, taken in July 2001, shows narrow sandy beaches and adjacent overwash sandflats, low vegetated dunes, and backbarrier marshes with ponds and channels. The second image shows the same site on August 31, 2005, two days after Hurricane Katrina made landfall on the Louisiana and Mississippi coastline. Storm surge and large waves from Hurricane Katrina submerged the islands, stripped sand from the beaches, and eroded large sections of the marsh. Today, few recognizable landforms are left on the Chandeleur Island chain. (Photo credit: NASA).

Fig. 8.8a,b The combined destructive forces of extreme wind velocities, record storm surge and up set waves from Hurricane Katrina (2005) left these signatures at single structures (a) and wiping out parts of settlements along the US Gulf coast near New Orleans (b) (Image credit: National Oceanic and Atmospheric Administration/Department of Commerce and National Coastal Data Development Center (NCDDC).

variability of these events. Proxy records collected from coastal environments offer the potential to extend this record back several thousand years, and may provide better statistical constraints on storm prediction and a better understanding of the influence of global climate change on catastrophic cyclone/hurricane development. During the last decade researchers studying the frequency and magnitude of ancient storms (the science of palaeotempestology) have compiled datasets based on different geologic archives with an appropriate memory. On the millennial to centennial timescale, archives encompass storm-induced geomorphic landforms; sedimentary and/or geochemical fingerprints in stratified sedimentary sequences, or with higher temporal resolution as isotopic traces in (biologic) archives such as speleothems, corals or tree-rings (see Emanuel, 2005; Knapp et al., 2010; Nott, 2004; Otvos, 2011; Scheffers et al., 2009, Travis, 2000). As the aim

Fig. 8.8c,d Hurricane Katrina in 2005 has left severe damage in the Mississippi delta region, USA. The Louisiana Superdome was converted into an island refuge for as many as 30,000 people. Many sheltered in the stadium while Hurricane Katrina raged overhead on Monday, August 29, 2005, and many more have been brought to the Superdome after being rescued from their flooded homes. The stadium itself was surrounded by flood water when the Quick Bird satellite captured this detailed image on August 31. Much of the stadium's bright white roof is missing. (Photo credit: NASA).

Data SIO, NOAA, U.S. Navy
Image © 2010 Terra
Image © 2010 Geo

Fig. 8.9 The signature of a late medieval winter storm land loss is still visible in the difference of land use patterns along the German North Sea coast. Areas of land loss and subsequent land reclamation are visible where the agricultural fields are much larger than the surrounding fields. This former bay had an extension along the coast of about 12 km. (Photo credit: ©Google Earth 2010).

Fig. 8.10 The Netherlands, an important industrially and densely populated nation, is situated on the delta of three of Europe's main rivers: the Rhine, the Meuse and the Scheldt with half of the country lying below sea level, including the most populated western part that contains Amsterdam and Rotterdam. Thus, it is not surprising that the establishment of protection measures against flooding is in a way a popular business in The Netherlands. Since the Middle Ages, extended dikes have been built to protect the country from high storm surges that push seawater inland. Behind the dikes millions of people are living and enormous investments have been made. (Photo credit: ©Google Earth 2010).

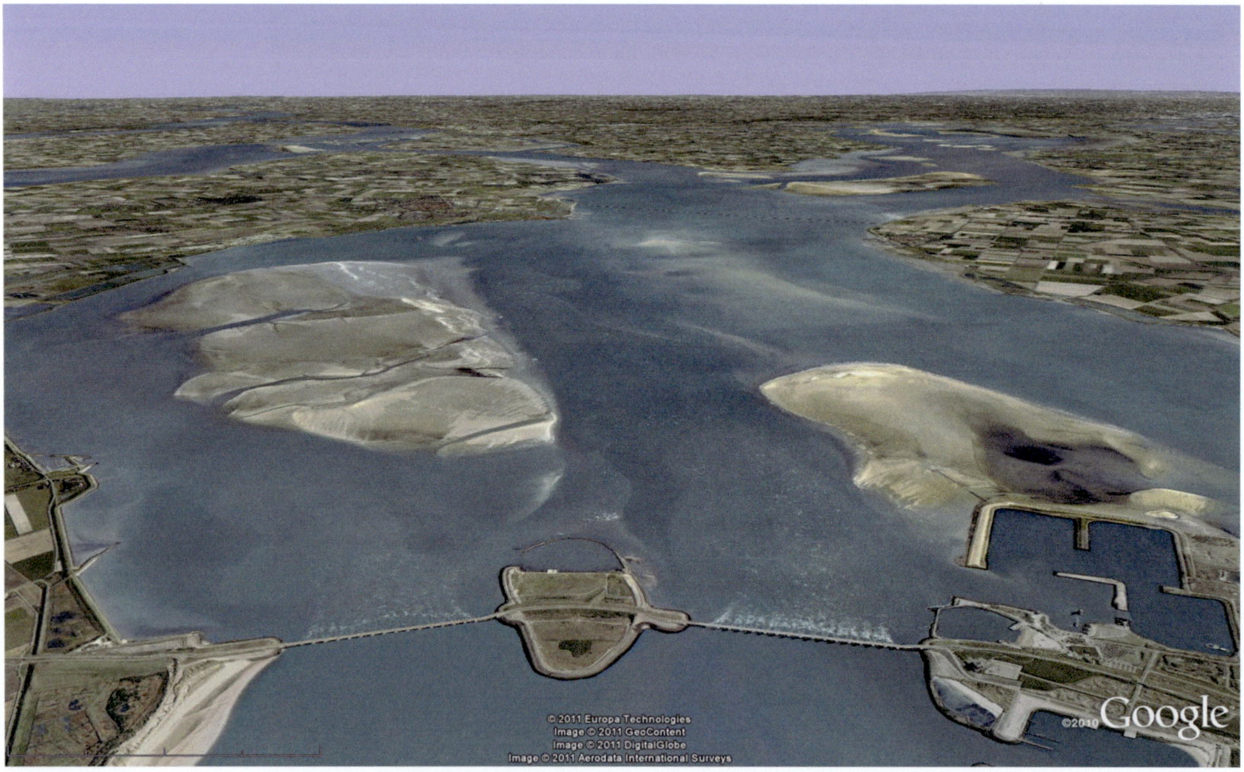

Fig. 8.11a

Fig. 8.11 a,b,c The Deltaplan protection works at the mouth of Scheldt river in the Netherlands protects the low-lying lands in the south western part of the Netherlands. The closing of the Eastern Scheldt Estuary was made a national priority after the catastrophic floods of February 1953. Originally, the Eastern Scheldt was to be a closed dam, but with increasing environmental awareness the Dutch started realising that only working with nature, and not against it, could ensure real safety. In 1975, the Dutch government proposed building an open barrier, instead of a simple closed dam, which could be closed during an emergency but would otherwise be open, preserving the ecology of the area. Now the Eastern Scheldt Dam has become quite an important tourist attraction, for visitors from all over the world. That makes sense, after all, the open barrier is one of the most impressive and beautiful maritime infrastructure feats ever conceived. The American Society of Civil Engineers (ASCE), a professional body founded in 1852, represents the members of the civil engineering profession around the world. This society made its own list of marvels of the civil engineering in an effort to recognize an actual equivalent to the Ancient Seven Wonders of the World. The Delta Work Plan of the Netherlands was listed a together with the Empire State Building, the Panama Canal, the Golden Gate Brigde or for example the Itaipu Dam. And the country has remained protected hitherto. (Photo credit: a) Google Earth 2010; b,c) istock).

of this book is to visualize forms and processes, a collection of coastal changes by storms is shown in the series of images below (Figs. 8.2–8.11).

Storm tides are also a regular occurrence in areas where winterstorms are common as is the case in the North Sea, here especially the Netherlands, northern Germany and Denmark are particularly at risk from storm surges. For the protection of the low-lying areas along the coast, long and high dike systems have been built over centuries (Fig. 8.9–8.11).

Tsunamis

Tsunamis are rather rare natural marine hazards, but their potential disastrous impact was shown to the world in a series of recent catastrophic events as during tragedy of the Indian Ocean Tsunami of December 26th, 2004 leaving in its wake about 225,000 fatalities and billions of US dollars in damage, or during the Fukushima tsunami in 2011, which caused the worst nuclear accident after Chernobyl in 1986 with fatal and unknown consequences far into the future. Tsunamis may be generated by displacement of the sea floor by submarine earthquakes and radiate from the epicentre of an earthquake. Tsunamis are not generated as a single wave – they consist of a wave-train. In the open ocean a typical tsunami may have a height of 0.4 m (and thus may not be perceived by ships at sea!), a wavelength of about 275 km, and travel between 600–800 km/h depending on water depth in the ocean basin. Tsunamis shoal in continental shelf waters due to friction with the ocean floor and can reach heights of over 40 m when they reach the shoreline as was witnessed by the world during the Indian Ocean Tsunami in 2004. Arriving at the coast tsunami waves usually not break but instead surge over foreshore. The height above sea level reached by a tsunami is called 'run-up'. Sometimes the sea will initially withdraw from the shoreline before the arrival of the wave. Because tsunami waves travel 10 times slower than seismic waves, there is enough time to warn distant shorelines of an impending disaster. The Pacific Tsunami Warning Center, based in Hawaii, rapidly locates oceanic earthquakes using the fast-moving seismic waves and estimates their potential for triggering a tsunami. A tsunami with fatalities is highly likely when the earthquake epicentre is shallow and the magnitude of the quake greater than magnitude 7 on the Richter scale. The Warning Center then quickly notifies the countries that may be in danger and an advance warning may be broadcast as much as a few hours before the arrival of the tsunami, allowing time for the evacuation of coastal populations. Unfortunately, no such system had been installed in the Indian Ocean, so the great 2004 tsunami struck with essentially no warning. Realizing how many tens

▶Fig. 8.12a,b The tsunami wave from the Fukushima earthquake (Japan) on March 11th, 2011. The epicentre of the magnitude 9.0 earthquake was located about 70 km off the coast at a depth of only 32 km. The quake occurred along about 500 km of the subduction zone between the Pacific Plate and the Honshu plate in the north-eastern part of Japan's largest island Honshu, close to the port-city of Sendai and about 380 km NE of Tokyo. The earthquake is hitherto the most powerful known to hit Japan and among the 5 most powerful earthquakes on Earth since start of measurements in 1900. The energy released was about 600 million times the energy of the Hiroshima atomic bomb. A part of NE Japan was pushed 2.4 m closer to the US with a subsidence of about 0.5 to more than 1 m, allowing the tsunami to travel faster and further inland. The seabed was uplifted in parts up to 3 m, and the horizontal movement of the sea floor near the epicentre was up to 24 m! The tsunami generated by this earthquake was devastating: The first wave hits the coast only 20 to 30 minutes after the earthquake with a wall of water and debris at most places towering more than 10 m high reaching a maximum of 37,9 m! On low lying coastal areas the water inundated the coast up to 10 km inland and covered at least 470 km^2 together with 101 tsunami evacuation sites. The number of fatalities including missing persons reaches a tragic number of 30.000 people. The impact was also felt on distant shores, even along the coastlines of Mexico and Chile houses have been destroyed by the tsunami wave. Although Japan invested in the past billions of US dollars in tsunami protection infrastructure with concrete sea-walls (16 m high), the devastation and destruction along the coast was extreme. About 190.000 houses have been destroyed or damaged, dozens of communities were wiped out by the tsunami and the livelihood of thousands of people annihilated. The staggering financial costs of Fukushima continue to rise as each week that passes brings new revelations about the unstable nature of the nuclear power plant facility at Fukushima as a result of the earthquake and tsunami. The catastrophe caused a nuclear accident and raises associations with two nuclear accidents in living memory: Three Mile Island in the US in 1979, and Chernobyl in Ukraine in 1986.The Fukushima nuclear disaster has rekindled the energy debate worldwide. And that is a good thing – after all, we cannot simply carry on as if nothing had happened after this event. We need strategies for controllable and sustainable power generation.

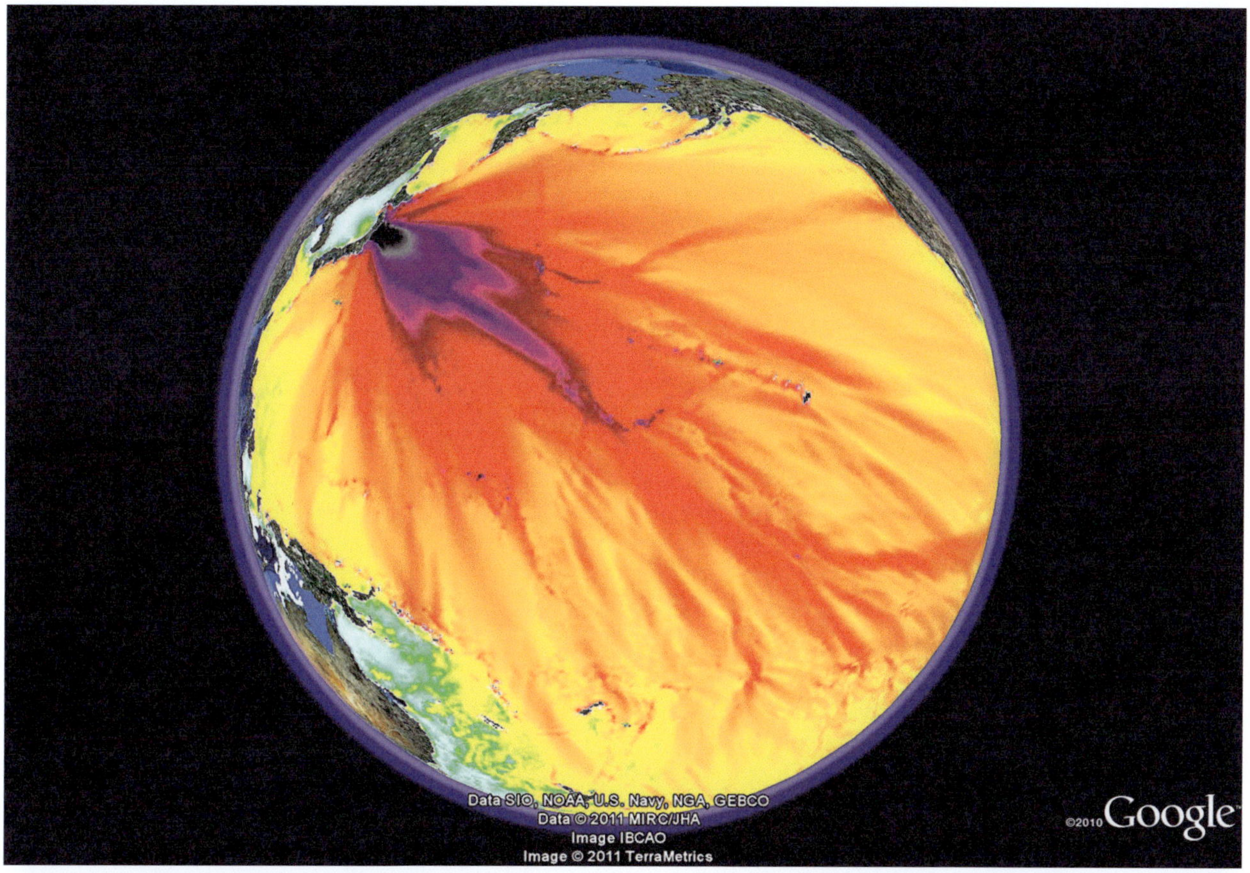
Data SIO, NOAA, U.S. Navy, NGA, GEBCO
Data © 2011 MIRC/JHA
Image IBCAO
Image © 2011 TerraMetrics
©2010 Google

8.12a

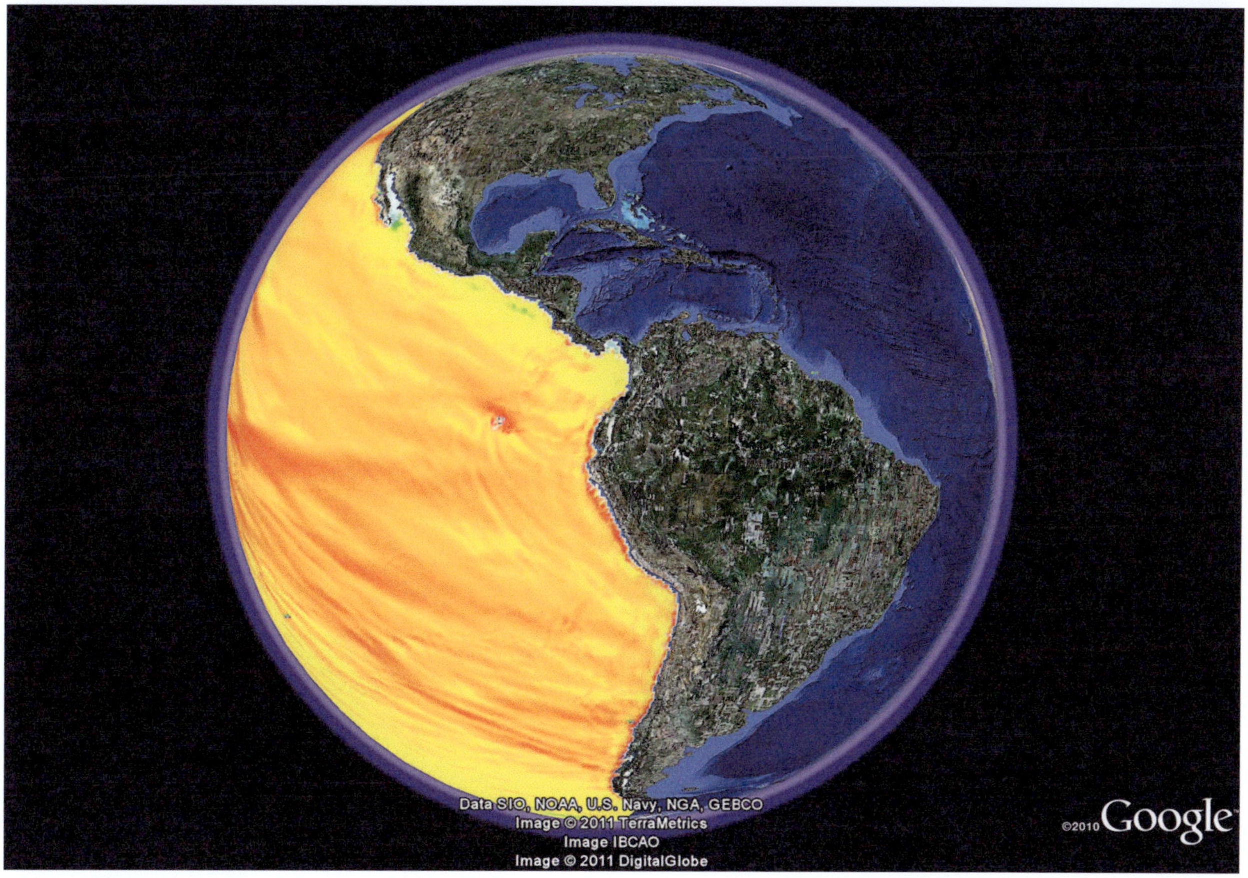
Data SIO, NOAA, U.S. Navy, NGA, GEBCO
Image © 2011 TerraMetrics
Image IBCAO
Image © 2011 DigitalGlobe
©2010 Google

8.12b

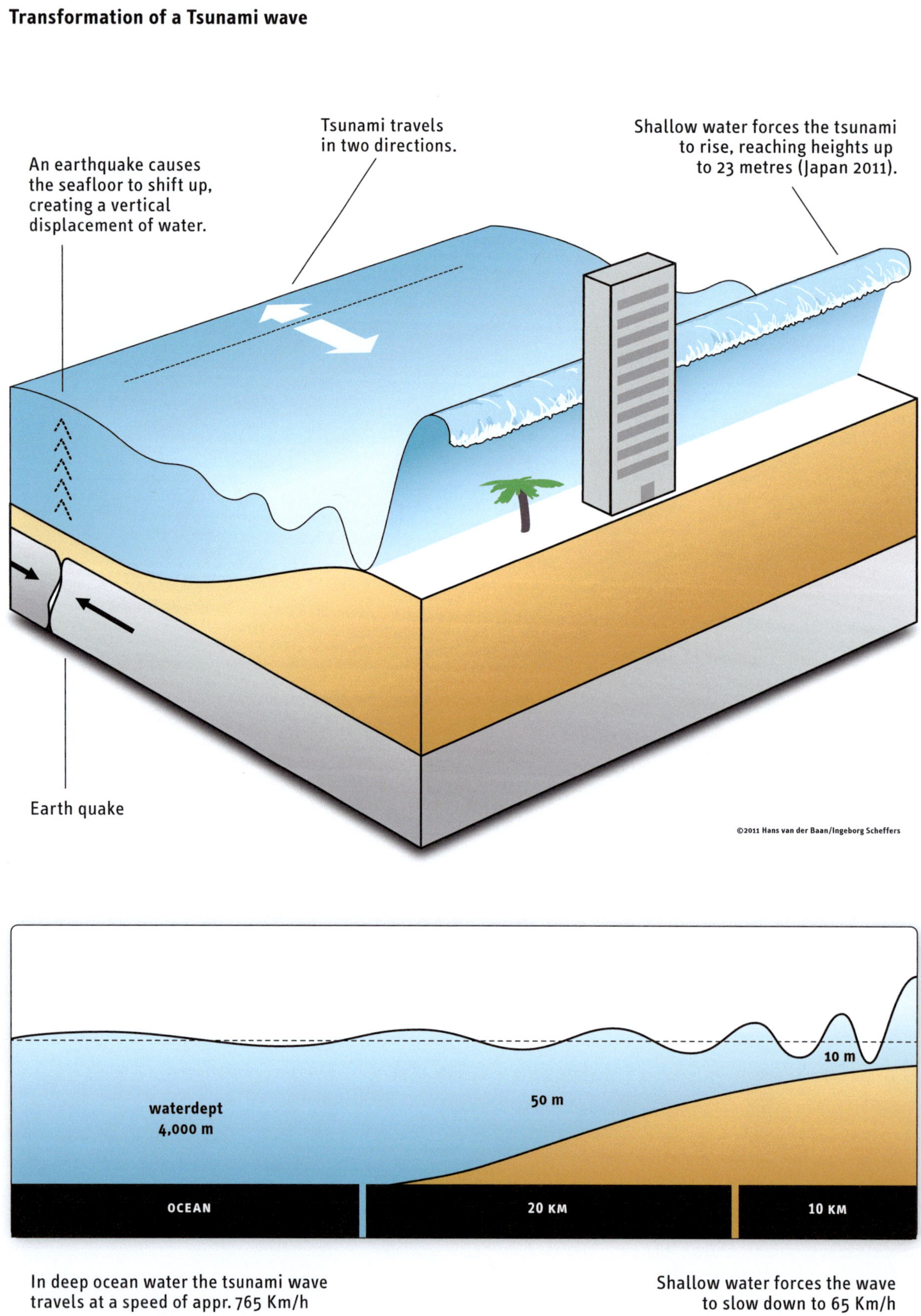

Fig. 8.13 Transformation of a tsunami wave approaching a shoreline.

The Storegga submarine slide mega-tsunami more than 7000 years ago in the Norwegian Sea and the places where its sediments have been found so far

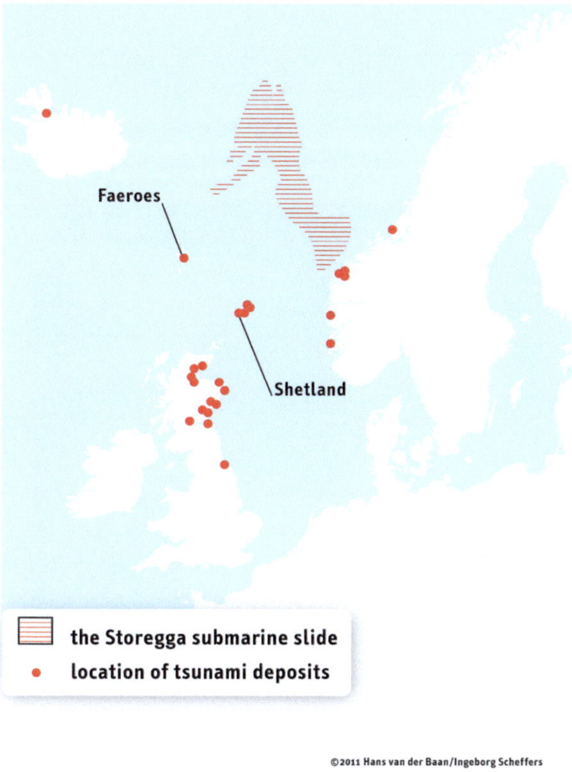

Fig. 8.14 The Storegga submarine slide mega-tsunami more than 7.000 years ago in the Norwegian Sea and the places where its sediments have been found so far (acc. to Smith et al., 2004, Weninger et al., 2008).

Submarine slides from around the Hawaian volcanic islands. Most of them are older than 100.000 years and have certainly triggered mega-tsunamis in the past.

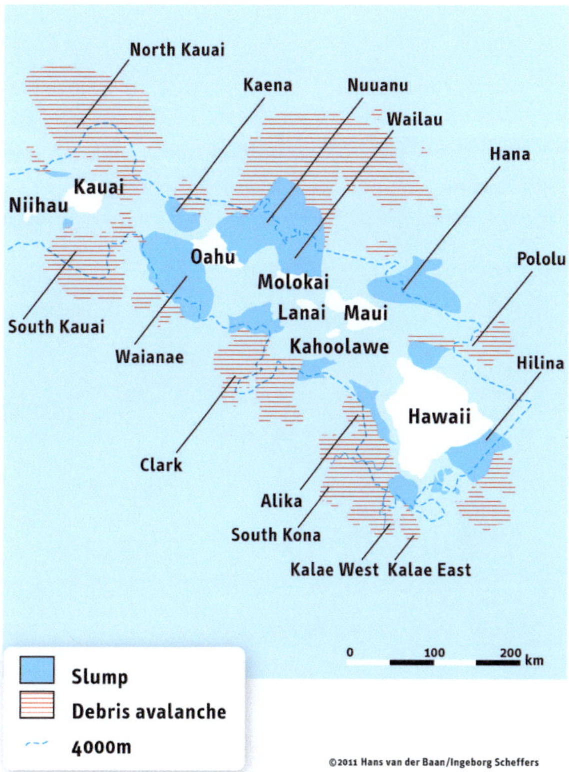

Fig. 8.15 Submarine slides of different types and sizes, many of them with several 1.000 km³ of mass around the Hawaiian volcanic islands. Most of them are older than 100.000 years and have certainly triggered mega-tsunamis in the past.

of thousands of lives could have been saved, nations around the world are now cooperating to develop tsunami warning systems for the Indian Ocean and to upgrade the existing systems in the Pacific and Atlantic oceans. The most difficult situation arises when a tsunami arrives so quickly that there is no time to warn nearby communities. This happened in the deadly 1998 Papua New Guinea tsunami, which killed as many as 3000 people in coastal villages near the epicentre. In this case, the best warning system is a very simple one: if you feel a strong earthquake, move quickly away from the coastal lowlands to higher ground!

Tsunamis may be triggered by several processes: The majority (more than 70%) are induced by undersea earthquakes, mostly in vicinity of active plate boundaries such as subduction zones. Other causes are underwater slides, land or large sections of ice slumping into the ocean, large volcanic eruptions or meteorite impacts in the ocean. Historians have compiled extended catalogues of tsunami events in human history, for the Mediterranean and China they date back to the last 4.000 years, but only a few of them have been identified and exactly dated by their deposits. As with storms, scientists try to extend the record of tsunamis into the past by studying and evaluating geologic evidence for ancient events to better understand the risks for the coastline's of the world (Figs. 8.12–8.24, and Benner et al., 2010; Bryant, 2008; Bryant & Nott, 2001; Imamura et al., 2008; Kelletat et al., 2007; Paris et al., 2005; Rhodes et al., 2006; Scheffers, 2006, 2008; Scheffers & Kelletat, 2003; Scheffers et al., 2008, 2009, 2010).

Fig. 8.16 Damage of the Tsunami of Dec. 26th, 2004, in Banda Aceh, Sumatra, Indonesia. (Photo credit: USGS; http://walrus.wr.usgs.gov/tsunami/sumatra05/).

Fig. 8.17 A sharp trimline along this island marks the flood depth with strong flow energy during the 2004 mega-tsunami at the coast of Sumatra. (Photo credit: ©USGS Tsunami Research Group).

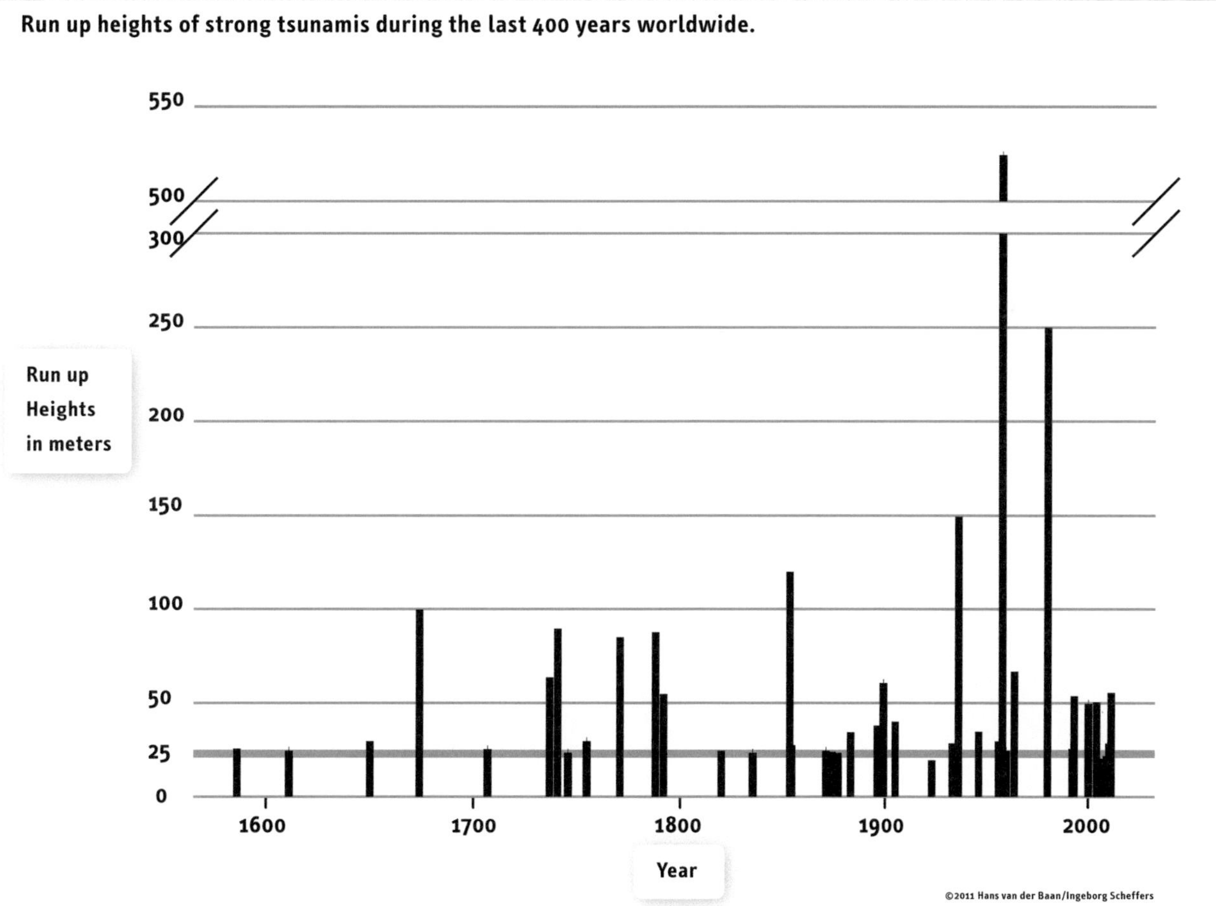

Fig. 8.18 Run up heights of strong tsunamis during the last 400 years worldwide. The maximum in 2004 along Sumatra was around +35 m, but we see that much higher levels have been reached, at least locally, with the maximum from a rock and ice fall in Lituya Bay in southern Alaska in 1958 with +525 m! (Credit: D. Kelletat / A. Scheffers).

Fig. 8.19a,b,c Locations in Banda Aceh at the northern part of Sumatra Island (Indonesia) immediately after the mega-tsunami of Dec. 26th, 2004. (Photo credit: ©Google Earth 2010).

Fig. 8.20 After the Indian Ocean Tsunami of 2004 new resorts have been rebuilt very close to the sea and are now undergoing severe erosion. Northern Thailand. (Photo credit: D. Kelletat).

Fig. 8.21 Tsunami damage in Japan in the 20th century. (Photo credit: USGS).

Fig. 8.22 A 16 m high massive concrete wall against tsunami impact at Tsu-shi, Japan. (Photo credit: Chris O at en.wikipedia)

Fig. 8.23 These two large boulders have been dislocated in one piece of about 170 tons more than 3.000 years ago on Bonaire island in the southern Caribbean. They have been broken during impact on the ground within the tsunami wave, about 160 m inland on a terrace +5 m above MHW. (Photo credit: A. Scheffers).

Fig. 8.24 One of the largest Holocene tsunami boulders worldwide with more than 2.000 tons on Tongatapu island in the Pacific Ocean (Photo credit: Matthew Hornbach / Cliff Frohlich, University of Texas; Frohlich et al., 2009, Geology 2009; 37;131–134).

8.2 Sea Level Rise – The unavoidable and uncertain future of our coasts

Since the onset of the Holocene warm period that followed the Last Ice Age around 10.000 years ago sea level has risen around 125 meters. This has natural causes due to eustatic sea level changes (climate related global changes due to the melting of large continental ice sheets and water mass added to the ocean) or isostatic effects by ice or sediment load and unload on the Earth's crust. Sea level rose relatively fast until around 6.000 years ago and since then has remained largely unchanged with fluctuations amounting to a few centimetres per century. The present sea level rise is a reaction to the average global warming of 0.8 degrees Celsius over the past 30 years (IPCC 2007), which is attributed to the human-induced greenhouse effect. Over the past two or three years, the science of climate change has become a more widely disputed issue in the public and political spheres, but recent observations of changes in the climate system strengthen the conclusions of the IPCC (Intergovernmental Panel on Climate Change) Fourth Assessment Report (2007) and the Garnaut Review (2008) that contemporary climate change is indeed real, and is occurring at a rapid rate compared with geological time scales (Climate Commission Secretariat, 2011). The evidence that the Earth's surface is warming rapidly is now exceptionally strong and beyond doubt. Sea level rise is one of the most series consequences of climate change. Since the 1880's when the first global estimates could be made, sea level has risen by about 20 cm. Satellite observations during the last two decades have measured a sea level rise of about 3.2 mm per year (1993–2009) compared to a rate of 1.7 mm per year for the 1900–2009 period (Church and White 2011). Admittedly, the measurements are for very short periods of time and so are difficult to extrapolate to longer time scales, but nonetheless they point to a substantial increase in the rate of sea level rise. A continuation of the currently observed rate of 3.2 mm/year would give a rise of about 0.32 m by 2100, about the mid-range of the IPCC scenarios. However, sea level is currently tracking near the upper range of the scenarios, and it seems unlikely that the rate of sea-level rise will remain fixed for nearly a century at its current level as the temperature continues to rise. The upper limit to sea level rise during this century as determined by the Intergovernmental Panel on Climate Change (2007) was placed at between 61 and 76 centimetres. The largest uncertainty in the projections of sea level rise is the behaviour of the large masses of ice on Greenland and Antarctica. A new study concludes that a "credible" maximum rise by 2100 would be about 2 meters but stresses that sea levels rising to the upper part of this range are unlikely (Nicholls et al., 2011). Using the IPCC's AR4 estimates the authors consider that a credible lower bound for the same time period would be 0.5 meters. Still, no one can really imagine how the coasts will look like if the waters rise by half a meter over the course of the century. Researchers try to assess the extent of threat posed by sea level rise for coastal areas by analysing current topographic heights above sea level to determine which regions are at the greatest risk of flooding. This may sound easy, but for many coastal areas no reliable topographic maps exist at the necessary resolution. Not surprisingly especially vulnerable even to small sea level rises are populated deltaic regions like Bangladesh or Vietnam, and many coastal cities and, should sea-level rise by around one meter, many island nation-states. The most geographically vulnerable regions – low-lying coastal regions as large parts of Asia and small islands – collectively contain the majority of potentially displaced peoples, but these are the same regions with low adaptive capacity and, therefore, where adequate protection measures will most likely not take place.

The example of the Netherlands shows that a small and affluent industrialized nation, when faced with a serious potential threat, is certainly capable of following the strategy of defence over the long term – after all, virtually two thirds of its country lies below the mean high-water mark. The Dutch even gear up for climate change with amphibious, floating houses that are firmly anchored on land like along the Maas dyke.

Venice, known as the "Queen of the Seas", serves as a dramatic example of the difficulties that can arise from human settlement too close to

Coastal areas most affected by sea-level rise

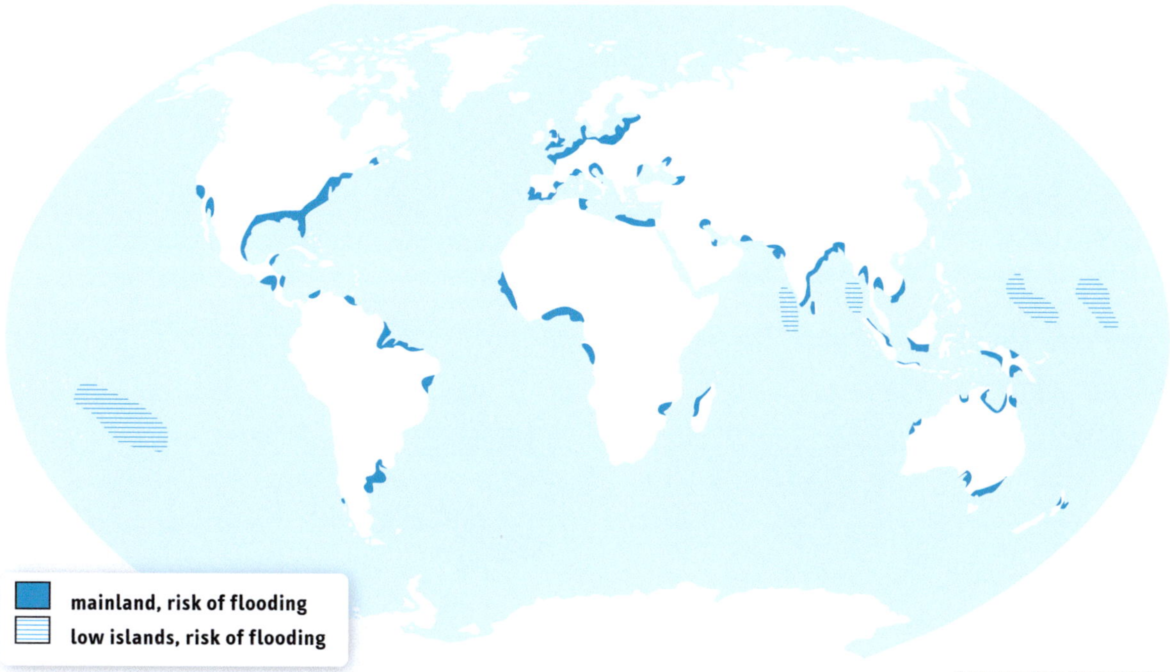

Fig. 8.25 Low-lying coastal areas are in particular at risk from sea level rise, floods, storm impacts and tsunami inundation (Image credit: Scheffers / Kelletat).

Top 10 nations classified by population of population in low-lying coastal regions			Top 10 nations classified by proportion of population in low-lying coastal regions		
Nation	Population 10^3	% of population	Nation	Population 10^3	% of population
1. China	127,038	10	1. Maladives	291	100
2. India	63,341	6	2. Bahamas	267	88
3. Bangladesh	53,111	39	3. Bahrain	501	78
4. Indonesia	41,807	20	4. Suriname	325	78
5. Vietnam	41,439	53	5. Netherlands	9,590	60
6. Japan	30,827	24	6. Macao	264	59
7. Egypt	24,411	36	7. Guyana	419	55
8. USA	23,279	8	8. Vietnam	41,439	53
9. Thailand	15,689	25	9. Djibouti	250	40
10. Philippines	15,122	20	10. Bangladesh	53,111	39

©2011 Hans van der Baan/Ingeborg Scheffers

Table 8.1 Nations with the largest populations and the highest proportions of population living in low-lying countries. (Not included countries with fewer than 100.000 inhabitants and 15 small islands states with a total of 423.000 residents. Source: World Ocean Report, 2010).

Fig. 8.26 Venice (NE Italy, at *26° 17′N* and *12° 19′E*) was built more than a thousand years ago on closely spaced oak tree pilons (mostly imported from today Slovenia and Croatia) in a lagoon. In the absence of sufficient oxygen under water, wood does not decay and becomes petrified as a result of the constant flow of mineral-rich water around and through it. (Photo credit: ©Google Earth 2010).

the coastal zone. Over 15 centuries ago a group of refugees from the Italian mainland, fleeing from northern invaders, settled on the islands located in the Venice Lagoon. The lagoon offered protection and easy access to important trade routes giving rise to the prosperity of the city. The Republic of Venice was a important centre of commerce (especially silk, grain and spice trade) and art in the 13th century up to the end of the 17th century and a major maritime power during the Middle Ages and Renaissance until the time of Napoleon. But the very same physical features that made Venice once so attractive for settlement now may bring the city to its end. Every year Venice loses a little land. Over the last century, through a combination of natural geological phenomena and anthropogenic influences like extraction of deep-well groundwater, Venice lost approximately 25 cm in height relative to sea level (Frassetto, 2005). The geological factors that contribute to the loss of land include subsidence and sea level change. Lagoons are complex, temporally changeable environments. Over the course of Venice's history, its inhabitants have attempted to shape the Lagoon environment to their needs. However, human intervention within the lagoon has frequently resulted in unanticipated consequences. Diversion of the rivers to prevent the lagoon from filling in has reduced the levels of sediments necessary for building up salt marsh beds to make up for loss from erosion and subsidence. Attempts to make the water ways within the lagoon more accessible to navigation have also accelerated the rates of erosion. The future of Venice seems tragic: Due to sea level rise, scientists project that even drastic intervention will only delay the permanent flooding of Venice 100–200 years.

Much poorer coastal and small island states are not in a position to protect their coastlines with sophisticated and expensive coastal defence works and are confronted with the choice of either adapting or retreating. But even resettlement projects for the first 1.700 climate change refugees from the Carteret Islands, part of Papua New Guinea, which began in 2007, are costly and amounting to several millions of US dollar.

Many of the risks due to sea-level rise are associated with inundation events from storm waves and surges or higher tides, which damage human settlements and infrastructure in low-lying coastal areas, and can cause considerable erosion of sandy beaches and sedimentary coastlines. While a sea-level rise of 0.5 m – less than the average waist height of an adult – may not seem like a matter for much concern, such modest levels of sea-level rise can lead to unexpectedly large increases in the frequency of extreme high sea-level events as such events are very sensitive to even small increases in sea level. A total of a billion people worldwide now live within 20 metres of mean sea level on land measuring about 8 million square kilometres. This is roughly equivalent to the area of Brazil. These figures alone illustrate how disastrous the loss of the coastal areas would be (World Ocean Report, 2010). Sea level rise will affect sandy coasts to a much larger extent compared to rocky coasts. Our ancestors adapted to risings seas by the adaption of buildings and settlements such as earth mounds or pile houses, more often human societies have chosen for retreat by abandoning and relocation of settlements until the Middle Ages. While the first strategy of adaption with innovative housing structures becomes more popular in urban planning like in the Netherlands, migration or orderly retreat today is the least accepted measure for coastal management scenarios. In modern times, the most applied strategy in North America, Europe, Australia and parts of East Asia is protection and defence by building dykes, sea walls or flood barriers, all a costly and complex engineering approach that poorer coastal countries can not adopt on a similar scale. For example, Germany and the Netherlands are spending about 250 million Euros for maintaining of already existing coastal defence structures. The choice of defence strategy depends not only on the costs but also on the geology of the area. For example, extensive dyke fortifications are simply too heavy and would sink in the soft sediments of the Ganges-Brahmaputra delta in Bangladesh.

The worldwide rise in sea level has not only dramatic effects in the foreseeable future, but contributed significantly to the fact that the foundations of prosperity of many coastal communities, sandy beaches, are eroding in most places. Since World War II a boom of coastal development, often carried out with a lack of foresight in construction or development planning, took place right at the waterfront. The negative repercussions of erosion and coastal retreat in terms of coastal protection have been an expensive exercise for many coastal communities. However, it is worth to quote the words of Finkl and Walker (2005) "The essence of the problem is not the dynamic adjustment of coastlines to fluctuating ambient conditions, but construction too close to the shore." In recent decades, more than 70% of the total length for the worlds' sandy beaches has retreated at a rate of at least 10 cm/year and 20% of the total length has retreated at a rate of more than 1 m/year. Any beach goer can observe geo-indicators that hint to sediment loss on beaches: Development of beach scarps in the berm, presence of tree stumps or marsh muds on the beachface, dune breaches with overwash fans or buildings precariously close to the surf zone are all signs that beaches move landward due to sediment loss. But what are the causes? A beach is an environment of constant movement and the sand budget with its input and outputs is maintained by sedimentation, erosion and transport. The beach will keep the same general form if the total sand input balances the total sand output – the beach is in dynamic equilibrium and appears stable, but is actually exchanging its material with the environments on all sides. If input and output are not balanced, the beach grows or shrinks and in cases, some coastal areas are stable over the long term because sediment is simply transported along them. The sediment that accumulates to form a beach may come from a variety of sources. In principal, any poorly consolidated material on which waves and currents may impinge can be a source. At any point along a beach, the beach gains sand from a number of sources – sand brought to the beach by waves, longshore currents and drift from nearshore and foreshore zones, material eroded from the backshore, are sediments carried to the shoreline by rivers or for example landslides. The beach also looses sediment in a number of ways – winds are transporting sand to the backshore dunes, longshore currents carry sand downdrift or storm waves remove sand in high-energy events. Especially dunes store important

Fig. 8.27 The Three Gorges Dam spans the Yangtze River in China and is the world's largest hydropower project in terms of installed capacity (20.300 MW). The dam flooded archaeological and cultural sites and displaced some 1.3 million people and attracts a lot of criticism in China and abroad. Located at *49° 24′N* and *111° 01′E*. (Photo credit: ©Google Earth 2010).

volumes of sand that help balance the budget of a beach. Anthony (2005) describes them as the "savings account" of the shore while the beaches act basically as a "checking account".

Causes of beach erosion are very complicated and often numerous factors are interrelated. Beside a relative rise in sea level natural factors include coastal uplift or subsidence, change in climate patterns (increasing storminess, deviation of prevailing wind systems) or changes of the littoral drift. One major cause of beach erosion, which is a worldwide occurring phenomenon, is the depletion of natural sediments (Bird, 1996). On many of the world's coasts, especially in areas where sea level over the past 5.000–6.000 years has been relatively stable, the sand forming the beaches was derived from sediments on the inner continental shelf. These drowned deposits have been reworked by waves and driven onshore to nourish beaches, often in form of extended beach ridges or dune systems (Anthony, 1995). This process of beach progradation has stopped in most regions as the nearshore sediment supply is depleted. Coastal land areas also grow where large deltas were formed over the past millennia. Rivers are an important transport path for sediments to be delivered to the coast. But in many parts of the world coasts are not replenished by sediments because natural river systems are altered by the construction of dams or other water reservoirs. Worldwide over 41.000 large dams are in operation and a multitude of smaller dams or reservoirs are impacting on stream flow and sediment transport. Together, they block 14% of the total global river flow and enormous volumes of sediment. This is a severe loss for the sediment budget of the coast leading to increased erosion. This sediment deficit is in particular problematic in places where the ocean floor is subsiding beneath the weight of sediment packages and new sediment accumulation is needed to compensate for subsidence as is the case in delta regions. The Nile River is a good example. After the construction of the Aswan Dam in the 1960's massive coastal erosion occurred. In the case of the Mississippi River, the sediment load today is about half what it was in pre-dam construction days (Finkl and Walker, 2005). A similar environmental nightmare is expected for the world's largest dam project – China's Three Gorges dam on the Yangtze River (Fig. 8.27).

8.3 Man-made Coastlines

Coastal erosion may also be introduced when beachfronts are developed for recreational, urban or industrial uses (Figs. 8.28–8.31). With multi-billion dollar assets under threat, beach erosion, its perception and remediation measures has become a controversial debated topic between coastal scientists, government agencies, local authorities and beachfront owners.

During the last century, many erosion control techniques used coastal armouring structures such as sea walls, breakwaters or jetties to reduce wave energy approaching the shore (Figs. 8.32–8.36) or to catch sediment moving along the shore and thus providing protection from coastal retreat. However, in some cases engineering work and artificial structures can exacerbate the problem: they were designed to cure or simply shifting it downdrift and causing leeward erosion (Figs. 8.37–8.41; Godschalk et al., 1989; Pilarczyk et al., 1996, Allsop, 2002, Reeve et al., 2004). Industrialized countries may be capable of holding back the sea.

Today, artificial beach nourishment is often an alternative and more environment friendly method of choice to protect sandy beaches from erosion and to combat coastal retreat (Fig. 8.41). The removal of sediment or sand from the nearshore zone, is a common procedure for beach nourishment projects along many coasts, but in the longer term may change the shape of the sea floor and thus the wave-break zone could shift to a position closer to shore (National Research Council, 1995; Dean, 2002).

Human activities also change the coastal zone through the tremendous growth of coastal cities, today 75% of the mega-cities with populations over 10 million residents are located in the coastal zone. Moreover, new, artificial land areas are encroaching into coastal zones like the airport of Nice (France) or Hong Kong or new real estate's for the very rich (Figs. 8.42–8.45).

Beaches are a multi-resource asset in many ways, both in developed and developing countries, but the combined effects of sea level rise and human

Fig. 8.28 Dredged channels as a waterway to a new marina in weakly cemented coral carbonates at the Florida Keys, USA (left) and the Palau islands of Micronesia (right). (Photo credit: D. Kelletat).

Figs. 8.29 New 5 star "overwater" hotels on Moorea and Bora Bora (French Polynesia), built into the lagoons of the fringing reefs. (Photo credit: D. Kelletat).

activity are pushing many coastal areas towards a crucial tipping point where any changes may become irreversible and a return to a sustainable, more natural equilibrium beach becomes more and more unlikely. We have seen that the shape of the coastal zone is governed by a balance of different factors including erosion, sedimentation, currents, tides and storm frequency and without doubt sea level will rise beyond the 21st century. An integrated coastal management approach should carefully take into account these factors with a thought to sustainability and protection of the coastlines of the world.

Fig. 8.30 Diamond mining on the beach and offshore has transformed long sections of the Namibia coastline at about *28° 31′S* and *16° 18′E,* reaching 5 km inland. (Photo credit: ©Google Earth 2010).

Fig. 8.31 Tourism is another cause of transformation of natural coastlines into artificial landscapes (Sinai peninsula at the Gulf of Aquaba, Egypt, at about *27°56′N* and *34°22′E,* in a coastal belt 2–4 km wide). (Photo credit: ©Google Earth 2010).

2010 DigitalGlobe
2010 TerraMetrics
age NASA
©2010 Google

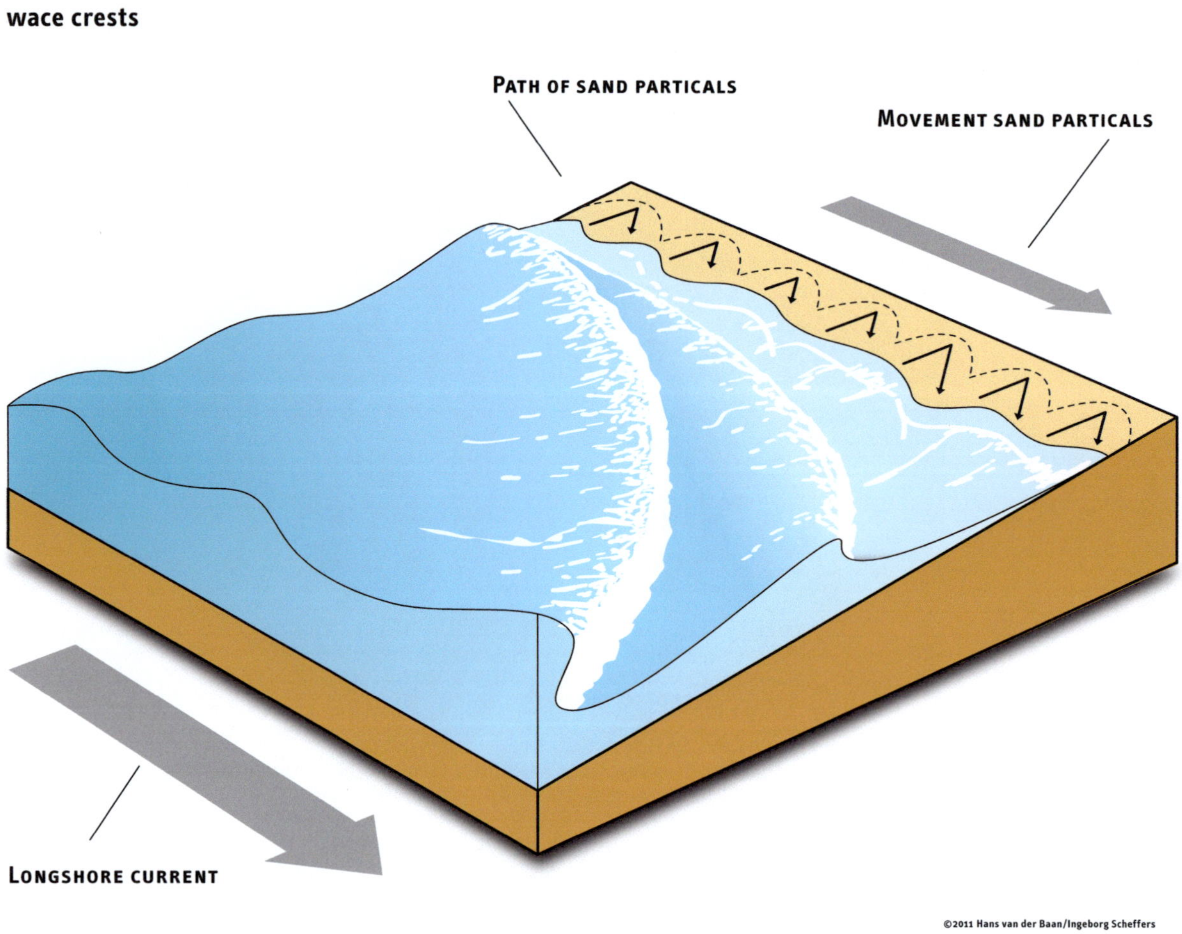

Fig. 8.32 The process of longshore drift by waves approaching with an angle to the coastline. If groynes or other obstacles are present, we normally find an accumulation at the luvward side and leeward erosion. If the coastlines set back, longshore drift may form a spit.

►Fig. 8.33 A coastline protected by groynes with a significant longshore drift: accretion of sand at the luvward sides and leeward erosion. West coast of Denmark at *56° 30′N* and *8° 06′E.* Section nearly 4 km long. (Image credit: ©Google Earth 2010).

►Fig. 8.34 Beach protection by concrete blocks as wave breakers along the Adriatic coast of Italy at *43° 44′N* and *13° 11′E,* scene width 2.5 km. (Image credit: ©Google Earth 2010).

Image © 2010 COWI A/S, DDO
Image NASA
Data SIO, NOAA, U.S. Navy, NGA, GEBCO
©2010 Google

Image © 2010 TerraMetrics
Image © 2010 GeoEye
©2010 Google

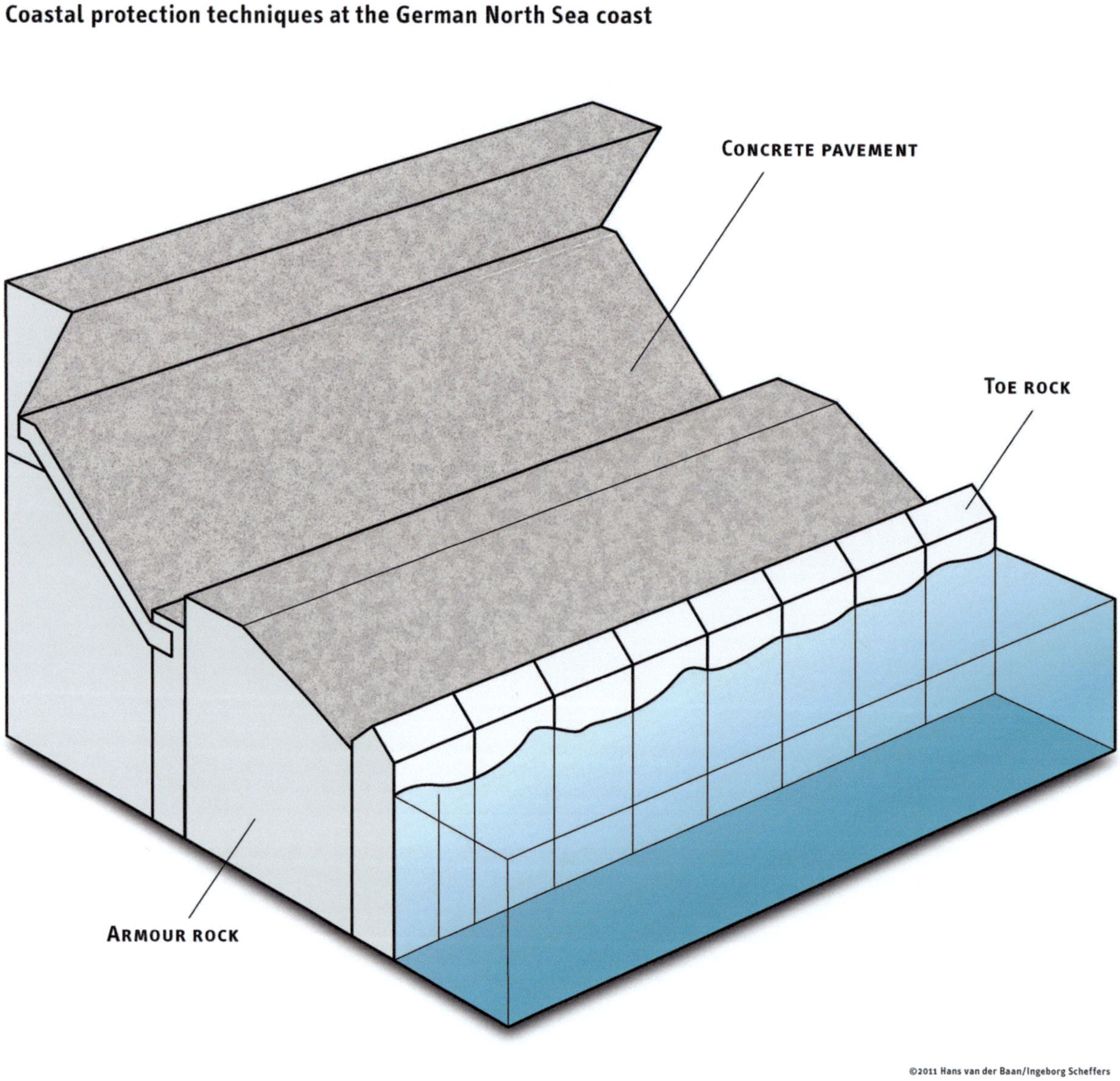

Fig. 8.35 Coastal protection techniques along the German North Sea coast.

Fig. 8.36 Concrete tetrapods as storm wave protection. (Photo credit: Istockphoto LP).

Fig. 8.37 a

Fig. 8.37 a,b Coastal protection with tetrapods changes sandy beaches into concrete constructions to combat ongoing beach erosion and sea level rise. Sylt Island, German North Sea coast. (Photo credit: a) D. Kelletat; b) ©Van Oord).

©2010 Google
Image © 2010 Aerodata International Surveys

8.38

Data SIO, NOAA, U.S. Navy, NGA, GEBCO
Image NASA
Image © 2010 Aerodata International Surveys
©2010 Google

8.39

Fig. 8.40 A strategy to stabilize dune cliffs with vegetation to catch and retain sediments along the outer coasts of the barrier islands of the German North Sea coast. (Photo credit: D. Kelletat).

◀ Fig. 8.38 High dikes form an artificial coastline in the Netherlands at *51° 27′N* and *3° 54′E*. Width of image is about 5 km. (Photo credit: ©Google Earth 2010).

◀ Fig. 8.39 Another example of dike shorelines of the Netherlands. (Photo credit: ©Google Earth 2010).

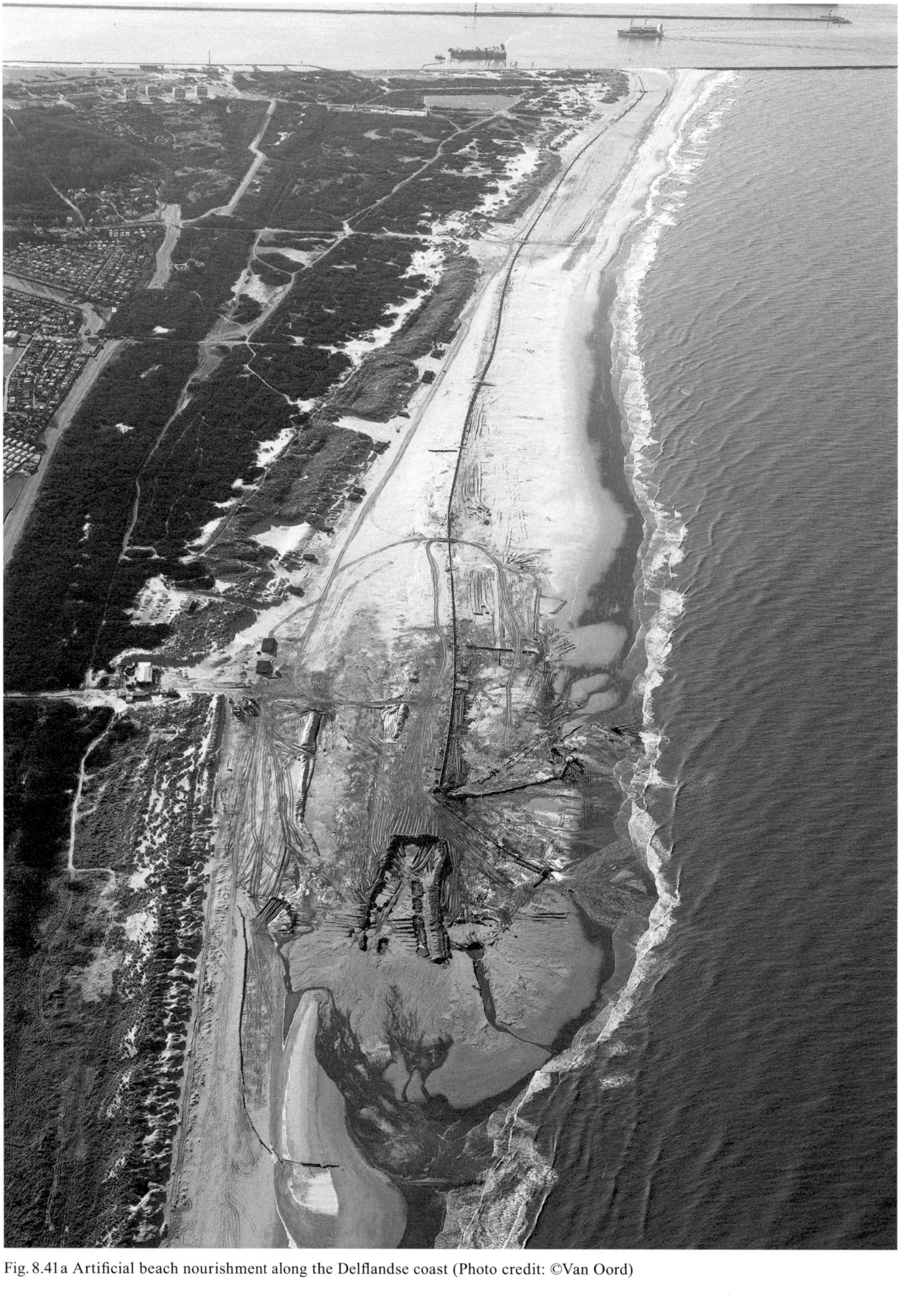

Fig. 8.41a Artificial beach nourishment along the Delflandse coast (Photo credit: ©Van Oord)

Fig. 8.41 b Artificial beach nourishment at the Tweede Maasvlakte, Rotterdam in The Netherlands (Photo credit: ©Van Oord)

Fig. 8.41 c,d Artificial beach nourishment of The World, Dubai (Photo credit: ©Van Oord)

Fig. 8.41e Artificial beach nourishment along Miami (Photo credit: ©USGS).

►Fig. 8.42 Land reclamation for city enlargement and industrial sites in the Bay of Tokyo, Japan. (Photo credit: ©Google Earth 2010).

►Fig. 8.43 A part of the land reclamation site near Tokyo, Japan. (Photo credit: ©Google Earth 2010).

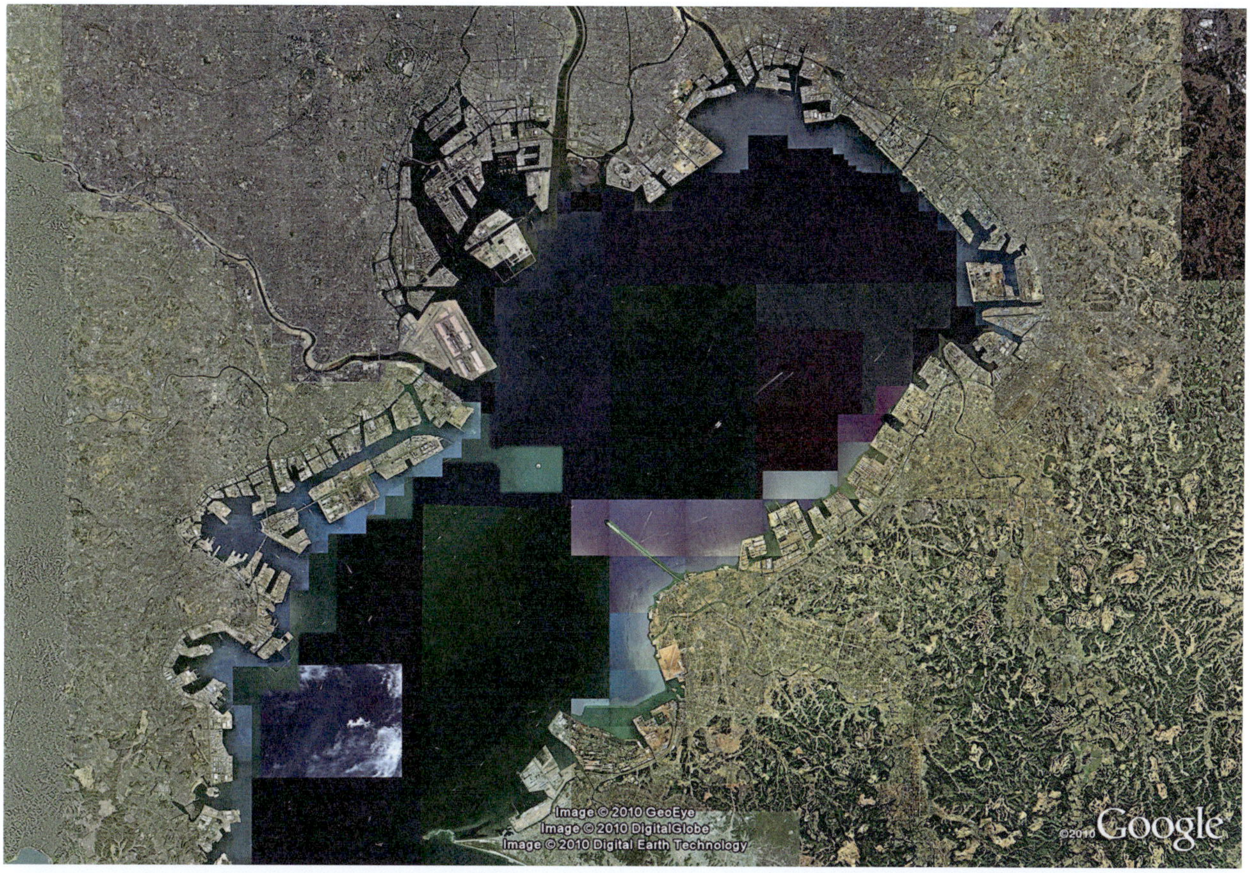
Image © 2010 GeoEye
Image © 2010 DigitalGlobe
Image © 2010 Digital Earth Technology
©2010 Google

8.42

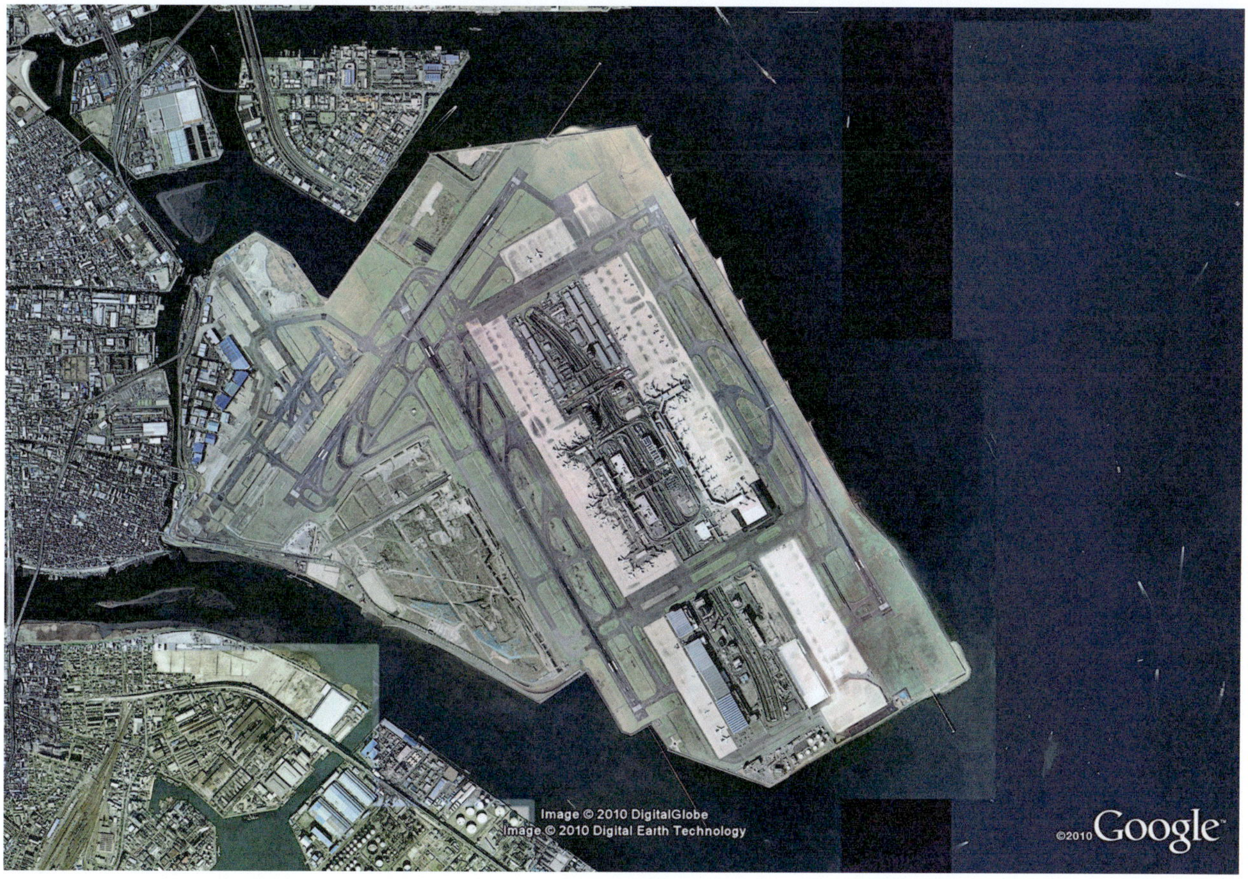
Image © 2010 DigitalGlobe
Image © 2010 Digital Earth Technology
©2010 Google

8.43

Fig. 8.44 a) Artificial islands (Palm Jumeirah) with about 5 km diameter at the Emirates coastline, *25° 07′N* and *55° 05′E*. (Photo credit: ©Google Earth 2010). b) Aerial photograph from Palm Jumeirah (Photo credit: ©Van Oord).

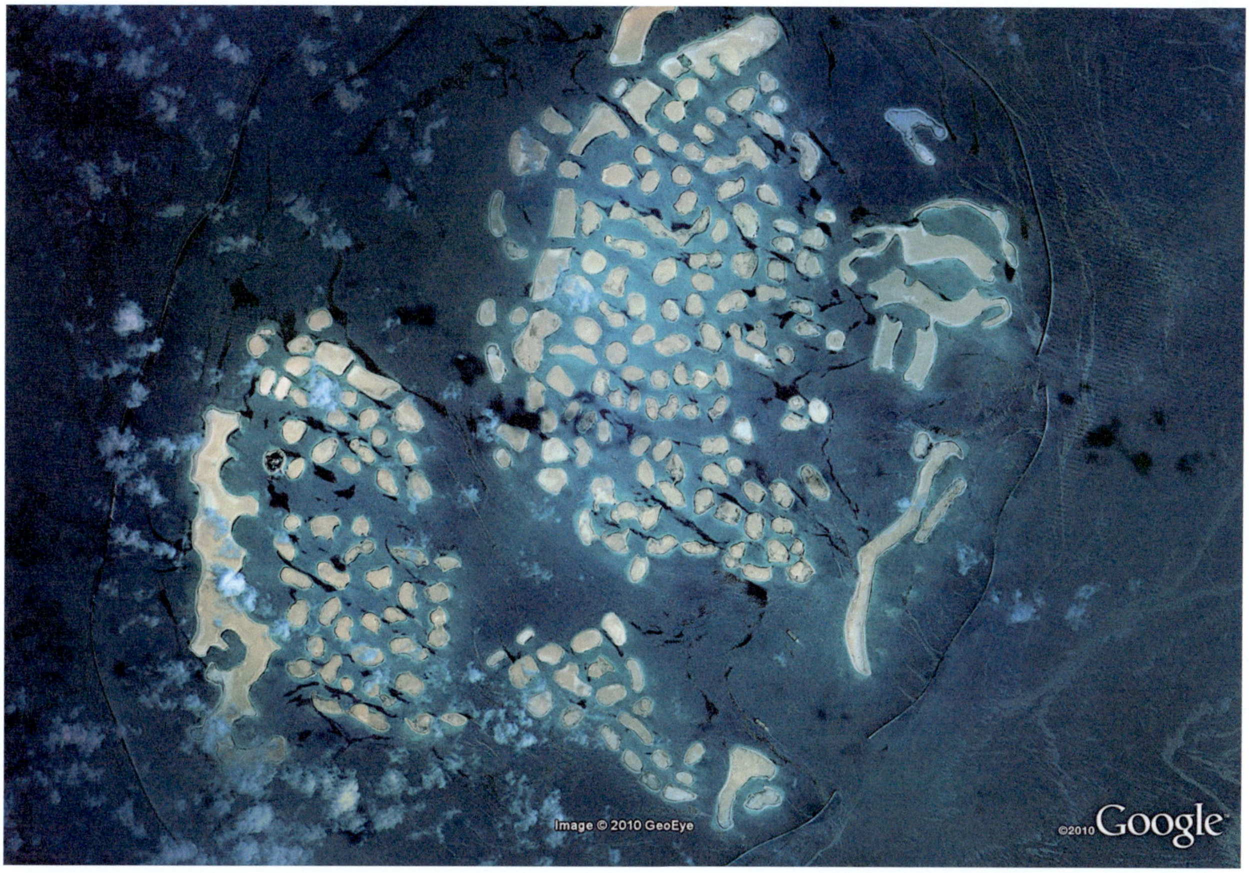

Fig. 8.45 Exclusive real estate developments on artificial islands in the form of the continents. The Emirates, Persian Gulf, at about *25° 13′N* and *55° 10′E*. "The World" has a diameter of 7,5 km. (Photo credit:A: ©Google Earth 2010, B, C: ©Van Oord).

References

Bryant E (2008) Tsunami – The Underrated Hazard, 2nd ed. Allsop, N. W. H. (2002) Breakwaters, coastal structures and coastlines: proceedings of the international conference organized by the Institution of Civil Engineers and held in London, UK on 26-28 September 2001. London: T. Telford.

Anthony, E. (2005) Beach Erosion. In: Schwartz, M. (Ed.) Encyclopedia of Coastal Science. Springer, Dordrecht.

Bird, E (1996) Beach management. Chichester. Wiley and Sons.

Bryant E (2008) Tsunami – The Underrated Hazard, 2nd ed. Springer, Dordrecht.

Church, J.A. and White, N.J. (2011) Sea level rise from the late 19th to the early 21st century. Survey of Geophysics. doi 10.1007/s10712-011-9119.1.

Climate Commission Secretariat, Department of Climate Change and Energy Efficiency, Commonwealth of Australia (2011) The critical decade. URL: www.climatecommission.gov.au ISBN: 978-1-921299-50-6 (pdf).

Dean, R.G. (2002) Beach Nourishment. Theory and Practice. World Scientific Publisher.

Emanuel KA (2005) Increasing destructiveness of tropical cyclones over the past 30 years. Nature 436:686–688.

Finkl, C and Walker, J. (2005) Beach nourishment. In: Schwartz, M. (Ed.) Encyclopedia of Coastal Science. Springer, Dordrecht

Frassetto, R. (2005) The facts of relative sea-level rise in Venice. New York, Cambridge University Press.

Garnaut, R. (2008) The Garnaut Climate Change Review. Final Report. Cambridge, UK: Cambridge University Press, 634 pp.

GESAMP (2008) Assessment and communication of environmental risks in coastal aquaculture. FAO Reports and Studies No 76, Rome.

Intergovernmental Panel on Climate Change (IPCC). (2007). Climate Change 2007: The Physical Science Basis. Contribution of Working Group I to the Fourth Assessment Report of the Intergovernmental Panel on Climate Change. Solomon, S., Qin, D., Manning, M., Chen, Z., Marquis, M., Averyt, K., Tignor, M.M.B., Miller, H.L. Jr and Chen, Z. (eds). Cambridge, UK and New York, NY, USA: Cambridge University Press.

Kelletat D, Scheffers A, Scheffers S (2007) Field signatures of the SE-Asian mega-tsunami along the west coast of Thailand compared to Holocene paleo-tsunami from the Atlantic region. Pure and Applied Geophysics 164 (2-3):413–431.

Knapp KR, Kruk MC, Levinson DH, Diamond HJ, Neumann CJ (2010) The International Best Track Archive for Climate Stewardship (IBTrACS): Unifying tropical cyclone best track data. Bulletin of the American Meteor. Society 91:3, 363–376.

National Research Council (1995) Beach Nourishment and Protection. Advanced Series in Ocean Engineering Vol. 18, National Academy Press, Washington D.C.

Nicholls, R.J. et al. (2011) Sea-level rise and its possible impacts given a 'beyond 4°C world' in the twenty-first century. Philosophical Transactions of the Royal Society A 369(1934): 161–181.

Nott J (2004) Palaeotempestology: The study of prehistoric tropical cyclones – a review and implications for hazard assessment. Environment International 30:433–447.

Otvos, EG (2011) Hurricane signatures and landforms – toward improved interpretations and global storm climate chronology, Sedimentary Geology,doi:10.1016/j.sedgeo.2011.04.014.

Paris R, Carrecedo JC, Torrado FJP (2005) Massive flanc failures and tsunamis in the Canary Islands: past, present, future. Annals of Geomorphology 140:37–54.

Pilarczyk, K. W.; Zeidler,& Ryszard (1996) Dutch case studies. Offshore breakwaters and shore evolution control. Taylor and Francis, London.

Reeve, D., Chadwick, A., and Fleming, C. (2004) Coastal Engineering. Processes, Theory, and Design Practice. Spon, USA.

Rhodes B, Tuttle M, Horton B, Doner L, Kelsey H, Nelson A, Cisternas M (2006) Paleotsunami Research. Eos 87 No 21:2, 205:209–210.

Royal Academy of Engineering (2011) Infrastructure, Engineering and Climate Change Adaptation – ensuring services in an uncertain future. ISBN 1-903496-61-6, London. Available online at www.raeng.org.uk/adaptation.

Scheffers A, Kelletat D (2003) Sedimentologic and Geomorphologic Tsunami Imprints Worldwide – A Review. Earth Science Reviews 63 (1-2):83–92.

Travis, J (2000) Hunting prehistoric hurricanes: Storm-tossed sand offers a record of ancient cyclones. Science News157: 333–335.

World Ocean Report (2010). Access under http://worldocean-review.com/en/.

Epilogue

Our long journey along the coastlines of the world has now come to an end. We hope you have enjoyed reading about coastlines and coastal phenomena within the 450 illustrations including 200 from a satellites' viewpoint, and maybe we have given some of you an inclination to have a closer look when you next enjoy a stroll on the beach. We also hope that the short descriptions throughout the book have been informative enough to better understand the extreme variety of forms and processes at coastlines, not only in terms of coastal hazards and risks but also with respect to their impressive beauty and aesthetic values. This is important, because only a better understanding and open minded inquisitive approach will allow us to protect the last untouched and unspoiled coastal landscapes of the world. And our coastal ecosystem are under severe threat, e.g. recent research has shown that an IPCC "business as usual" scenario of carbon emissions will result in catastrophic coral reef decline worldwide possibly already within the next decade!

In the 21st century there is a delicate balance between ever increasing numbers of people using coastlines for recreation and business yet, this also increases the chance that natural disasters transform into national catastrophes for individuals and societies. And, for the first time in the history of our planet, mankind influences natural processes on a global scale as for example sea level rise or endangering coastal ecosystems. Our technical and scientific capabilities enable us to safeguard our planet with sustainable management, only if we all agree to the importance of conserving our coasts. We are convinced that it is not too late to start preserving our planet and its precious resources, and if this book contributed to protecting our coastlines' beauty and variety, we have reached our goal.

Index

W

Z

MIX
Papier aus verantwortungsvollen Quellen
Paper from responsible sources
FSC® C105338

If you have any concerns about our products,
you can contact us on
ProductSafety@springernature.com

In case Publisher is established outside the EU,
the EU authorized representative is:
Springer Nature Customer Service Center GmbH
Europaplatz 3, 69115 Heidelberg, Germany

Printed by Libri Plureos GmbH
in Hamburg, Germany